Mobile Genetic Elements

Frontiers in Molecular Biology

SERIES EDITORS

B. D. Hames
Department of Biochemistry and Molecular Biology
University of Leeds, Leeds LS2 9JT, UK

AND

D. M. Glover
Cancer Research Campaign Laboratories, Department of Anatomy and Physiology
University of Dundee, Dundee DD1 4HN, UK

TITLES IN THE SERIES

---

1. **Human Retroviruses**
Bryan R. Cullen

2. **Steroid Hormone Action**
Malcolm G. Parker

3. **Mechanisms of Protein Folding**
Roger H. Pain

4. **Molecular Glycobiology**
Minoru Fukuda and Ole Hindsgaul

5. **Protein Kinases**
Jim Woodgett

6. **RNA–Protein Interactions**
Kyoshi Nagai and Iain Mattaj

7. **DNA–Protein: Structural Interactions**
David M. J. Lilley

8. **Mobile Genetic Elements**
David J. Sherratt

9. **Chromatin Structure and Gene Expression**
Sarah C. R. Elgin

10. **Cell Cycle Control**
Chris Hutchison and David M. Glover

11. **Molecular Immunology** (Second Edn)
B. David Hames and David M. Glover

Mobile Genetic Elements

EDITED BY

David J. Sherratt

Microbiology Unit,
University of Oxford

OXFORD UNIVERSITY PRESS
Oxford New York Tokyo

Oxford University Press, Walton Street, Oxford OX2 6DP

Oxford New York
Athens Auckland Bangkok Bombay
Calcutta Cape Town Dar es Salaam Delhi
Florence Hong Kong Istanbul Karachi
Kuala Lumpur Madras Madrid Melbourne
Mexico City Nairobi Paris Singapore
Taipei Tokyo Toronto

and associated companies in
Berlin Ibadan

Oxford is a trade mark of Oxford University Press

Published in the United States
by Oxford University Press Inc., New York

A catalogue record for this book is available from the British Library

Library of Congress Cataloging in Publication Data
Mobile Genetic elements / edited by David J. Sherratt.
(Frontiers in molecular biology)
Includes bibliographical references and index.
1. Mobile genetic elements. I. Sherratt, David J. II. Series.
QH452.3.M64 1995 547.87'322—dc20 94–23799

ISBN 0 19 963405 X (Hbk)
ISBN 0 19 963404 1 (Pbk)

Typeset by
Footnote Graphics, Warminster, Wilts.
Printed in Great Britain by
The Bath Press, Avon

Preface

Mobile genetic elements are all-pervading; they inhabit our chromosomes; are major sources of genetic plasticity; and are exploited as major genetic tools in many areas of biological research. Jim Shapiro introduces the volume with his very personal account of the discovery and significance of mobile genetic elements. He describes the pioneering contributions of Barbara McClintock's work on 'controlling elements' in Maize during the fifteen years that preceded the determination of the structure of DNA. Shapiro then goes on to show how in the 1960s and beyond, molecular analysis provided physical reality to these elements, and provided insight into the interplay of these elements and the hosts they inhabit. Early analyses provided descriptions of many different, apparently unrelated, elements in different organisms. These elements shared little DNA similarity and appeared to show a myriad of different transposition mechanisms. A triumph of mechanistic analysis has subsequently shown that the chemistry of transposition is conserved from retroviruses to bacterial transposable elements. This mechanism and its consequences is discussed in Ron Plasterk's chapter, while Marshall Stark describes processes and mechanisms that provide selectivity in the rearrangement process, using as his main example site-specific recombination reactions. Transposable elements have become major genetic tools in many organisms. They can be used as genetic mutagens, as 'reporters' of gene expression, and for isolating genes and for manipulating chromosome structure. Chapters by Claire and Doug Berg, and by Kim Kaiser, John Sentry, and David Finnegan describe how bacterial and eukaryote transposable elements can be used as powerful genetic tools. Not all mobile genetic elements are classical transposable elements. We now have integrons, shufflons, mobile introns, and retrons. Surely there will be more to come. Though it has not been possible to cover all of these new elements, the chapter on bacterial retrons by Dongbin Lim, Tania Lima, and Werner Maas gives a flavour for the diversity of elements and shows that reverse transcriptase is not confined to eukaryotes. Where do such elements originate? There is now substantial evidence that the mobile elements of all types have crossed species boundaries, have infected new populations, and may become extinct. Like viruses they are generally 'selfish', the process of mobility after leading directly or indirectly to element duplication. Nevertheless, much of the time they hide within genomes where they can exert subtle as well as dramatic effects on cell and organism biology. John Brookfield discusses these population and evolutionary aspects of mobile genetic elements.

My thanks go to all of the authors for their enthusiasm for this project and for their attempts to provide new and accessible insight into the properties and exploitation of mobile genetic elements. We hope that this volume will provide stimulating reading and useful information to established researchers from many disciplines and to young scientists who have yet to decide on their future careers.

Oxford, UK David J. Sherratt
November 1994

Contents

Contributors

CLAIRE M. BERG
U-131, Department of Molecular and Cell Biology, The University of Connecticut, Storrs, CT 06268, USA.

DOUGLAS E. BERG
Box 8230, Department of Molecular Microbiology, Washington University Medical School, St Louis, MO 63110, USA.

MARTIN R. BOOCOCK
Institute of Genetics, University of Glasgow, Church Street, Glasgow G11 5JS, UK.

J. F. Y. BROOKFIELD
Department of Genetics, School of Biological Sciences, University of Nottingham, Nottingham NG7 2UH, UK.

DAVID J. FINNEGAN
Institute of Cell and Molecular Biology, University of Edinburgh, Edinburgh EH9 3JR, UK.

KIM KAISER
Institute of Genetics, University of Glasgow, Glasgow G11 5JS, UK.

DONGBIN LIM
Department of Microbiology, Gyeongsang National University, Chinju 660-701, South Korea.

TANIA M. O. LIMA
Department of Microbiology, New York University Medical Center, New York, NY 10016, USA.

WERNER K. MAAS
Department of Microbiology, New York University Medical Center, New York, NY 10016, USA.

RONALD H. A. PLASTERK
Division of Molecular Biology, The Netherlands Cancer Institute, Plesmanlaan 121, 1066 CX Amsterdam, The Netherlands.

JOHN W. SENTRY
Institute of Genetics, University of Glasgow, Glasgow G11 5JS, UK.

JAMES A. SHAPIRO
Department of Biochemistry and Molecular Biology, University of Chicago, 920 E. 58th Street, Chicago, IL 60637, USA.

W. MARSHALL STARK
Institute of Genetics, University of Glasgow, Church Street, Glasgow G11 5JS, UK.

1 | The discovery and significance of mobile genetic elements

JAMES A. SHAPIRO

1. Introduction

One of the most intriguing facts about the history of mobile genetic elements is that no-one set out to discover them. In every case, they were initially found by geneticists studying other problems: chromosome breakage and rejoining in maize, transfer of genetic information between bacterial cells, lysogeny, regulation of gene expression, or antibiotic resistance. Is there a message in this repetitive serendipity? I think so. The message appears to concern the power that unspoken assumptions hold over our thinking (in this case, the idea that the genetic material is basically stable) and the need to be alert for those unexpected findings which overturn those assumptions and suddenly open a whole new universe for scientific exploration. Such major reorientations in thinking are inevitable in science (1).

Genetics has travelled a long conceptual road from the Constant Genome paradigm of the 1940s, when McClintock first described transposable controlling elements in maize, to the Fluid Genome paradigm that reigns today. There is an even longer and more fascinating voyage ahead as we come to terms with the implications of mobile genetic elements and molecular discoveries about genome organization and reorganization. This historical chapter affords a rare opportunity to trace the outlines of the intellectual journey already completed as well as to attempt some projections of where we might be going. The emphasis here will be on new ways of looking at genetic problems, on key observations, and on the unexpected convergence of discoveries in separate experimental systems. Accounts of the history of particular systems with a more mechanistic focus can be found in the first *Mobile Genetic Elements* book (2).

2. Maize controlling elements

Barbara McClintock discovered transposable elements in maize after many years of studying chromosome breakage and its genetic consequences (3). Her work on chromosome breakage began by investigating genetic instabilities associated with X-ray-induced chromosome rearrangements, especially ring chromosomes (4). She had found that nuclei harbouring two broken chromosome ends will join them together quite rapidly and efficiently. Depending upon the positions of the breaks, these joining events created a wide variety of chromosome abnormalities, including translocations, inversions, duplications, deficiencies (what we now call deletions), and chromosome losses. McClintock had learned to follow the loss of specific markers as chromosome changes occurred during development by observing variegated patterns of pigmentation in maize plants and kernels.

In the early 1940s, McClintock set out to investigate the genetic content of a single chromosome arm. She had devised a system to initiate breakage in the short arm of chromosome 9 during meiosis. Self-fertilization of plants carrying these broken chromosomes would lead to the formation of embryos undergoing what McClintock termed the chromosomal type of breakage–fusion–bridge cycle (see the papers in reference 3 for a full account). Plants emerging from zygotes subjected to the breakage–fusion–bridge cycle were expected to contain modified chromosome 9s, and they did. But they also displayed new kinds of genetic instability, which McClintock had not observed before. Instead of the sporadic patterns of chromosome loss she was familiar with from her breakage studies, the new instabilities showed certain regularities which told her that a controlled process was taking place. She spent the next three decades documenting the genetic elements which underlay these instabilities and uncovered an unsuspected array of genomic systems capable of changing chromosome structure and altering the regulation of gene expression. Because of their capacity for creating different patterns of gene expression during development, McClintock applied the name 'controlling elements' to the newly discovered mobile genetic systems (5).

One of the novel types of genetic instability McClintock found was manifested as repeated breakage events at a specific chromosomal locus. McClintock called these specific breakages 'dissociation' and labelled the responsible element at the breakage site *Ds*. The breakages could be followed by observing clonal patches of tissue (sectors) in either maize plants or kernels in which dominant markers distal to *Ds* on the chromosome had been lost. How many such clones were formed and when they arose during plant or kernel development was often quite regular. In other words, the frequency and timing of chromosome breaks was non-random and under some kind of control. The control system could be reset, however, and when this resetting happened, sectors or plants would appear in which a new frequency/timing regime was expressed. The *Ds* system was also quite regular in that the same group of markers was lost, and physical breaks could be seen through the microscope to occur at the appropriate position in chromosome preparations (6). Occasionally, however, a plant sector or kernel was observed to change pattern such that a new group of markers was lost in each sectoring event. Such results indicated that *Ds* had changed its position, and this new position could be confirmed by cytological examination. Thus, *Ds* was found to be capable of transposition from one site to another.

Genetic crosses quickly revealed that regulated chromosome breakage did not depend only on the *Ds* element at the site of breakage events. Chromosomes carrying *Ds* could be quite stable in some plants, but they would again show breakage when introduced into plants that carried another genetic element, termed *Ac* for activator of *Ds*. Like *Ds*, *Ac* could transpose from one location to another. The timing and frequency of *Ds* breakage events was also regulated by *Ac*: the higher the *Ac* dosage in a tissue, the later *Ds* breaks occurred during development. This *Ac–Ds* system came to be the prototype for what are now called 'two-element systems', where the activity of one dependent transposable element is conditional upon functions encoded by another autonomous element. Other examples of two-element systems include the Dotted (*Dt*) and Suppressor–Mutator (*Spm*) autonomous elements, each of which has its corresponding family of dependent elements. *Ac* does not activate *Spm*- or *Dt*-responsive elements, and neither *Spm* nor *Dt* will induce genetic instability at *Ds*. These maize systems were, incidentally, the first examples demonstrating that specific gene products encoded by one locus could regulate the behaviour of another locus, and McClintock drew attention to some parallels of controlling element functions with the kinds of regulatory molecules described by Jacob and Monod (7).

The second novel type of genetic instability coming from the self-fertilized breakage–fusion–bridge cycle plants involved 'mutable loci' (8). These were mutant loci affecting some visible phenotype, such as pigmentation, which gave rise to variegation patterns showing changes from the recessive to the dominant phenotype during development. For example, clonal patches of pigmented cells would appear on a mutant, unpigmented aleurone layer of a kernel. These mutable loci could affect many different characters, such as chlorophyll and anthocyanin pigmentation or endosperm starch synthesis, and mutable alleles could be found at any known locus where a suitable selection scheme was applicable (9). Superficially, the recessive to dominant variegation at mutable loci was the opposite of the patterns given by chromosome breakage, where marker loss gave a change from dominant to recessive phenotype. But the underlying similarity between genetic events at *Ds* and at mutable loci became manifest when it was found that the presence of *Ac* was necessary for instability to appear at a number of mutable loci. Ultimately, it became clear that some mutable alleles represented transpositions of *Ds* elements into particular loci and that recovery of dominant expression was the result of *Ac*-dependent *Ds* excision from the locus. Other mutable alleles resulted from the insertion of *Ac* itself or of controlling elements belonging to the *Spm* or other controlling element families (10).

One feature of mutable alleles which needs to be mentioned is the variety of expression patterns they generated at any given locus. Variegation was only one kind of novel pattern. Quantitative changes in stable expression could be observed in the absence of the mutator (transposase) function of the autonomous element, and a number of mutable alleles gave rise to different levels of expression following element excision (10). Other changes affected tissue specificity and spatial distribution of expression in the absence of variegation. McClintock derived variants of anthocyanin pigmentation loci which encoded kernel patterns similar to those treasured by Amerindians for religious purposes (10). In addition, she has indicated that controlling elements can be used to analyse the genetic basis of non-clonal patterns, where groups of cells that are not related to each other by descent display common phenotypes (11, 12).

After a decade of intensive genetic and molecular analysis, we are beginning to have a detailed picture of how controlling elements move through the maize genome and of the myriad mechanisms by which they alter the expression of information encoded at specific loci (13). Even without the molecular details, however, McClintock's discoveries had revolutionary implications. First, of course, was overturning the Constant Genome notion. Her work showed that cells contained activatable systems that could restructure the genome. Once recombinant DNA methods made all organisms amenable to molecular genetic analysis, the universality of mobile genetic elements and the diversity of their molecular mechanisms became clear. Today we think of the Fluid Genome.

A second fundamental consequence of McClintock's work concerned the concept of the unitary gene. If controlling elements could insert into individual loci and alter their pattern of expression during development, then each genetic locus could not be occupied by a series of indivisible alternative alleles. Instead, a genetic locus must be a complex mosaic structure that can be modified by the addition and removal of specific genetic elements. A related third point to emerge from McClintock's work was the realization that individual loci are not autonomous units but can be connected into a co-ordinately controlled system. When members of the same controlling element family inserted into two loci, expression of those hitherto separate loci came under joint control of the cognate autonomous element. Simple examples like these illustrated how repetitive genetic elements could be distributed to many sites in the genome to constitute the physical basis for an intranuclear regulatory network (14).

3. Plasmids, phages, and episomes

The second chapter in the history of mobile genetic elements began in the late 1940s with the effort to understand two of the mechanisms of genetic exchange in bacteria: cell-to-cell conjugation and bacteriophage-mediated transduction. The serendipity theme has been very strong in bacterial genetics. Bill Hayes discovered the F plasmid as part of a long-term effort to understand phase variation. Elie Wollman and François Jacob discovered F insertion in the circular bacterial chromosome as a consequence of experiments designed to map the λ prophage. Alan Campbell figured out reciprocal recombination in prophage insertion and excision as the consequence of mapping defective λ transducing phages. No-one anticipated that the concept of episomes (mobile genetic elements which can exist in two alternative states, either autonomous or attached to the bacterial chromosome) would grow out of studying the distinct phenomena of genetic recombination and lysogeny.

Hayes' experiments were technically simple but conceptually elegant. Following the discovery of *Escherichia coli* recombination by Lederberg and Tatum (15), Hayes used minimal medium to detect the formation of recombinants between auxotrophic mutants. His innovation was to use streptomycin as a selective agent in his crosses (16). Because streptomycin kills potential recombinants before they can form colonies, Hayes could distinguish the bacteria in a fertile cross into donors and recipients. Streptomycin-sensitive donors and streptomycin-resistant recipients would form recombinant colonies on medium containing streptomycin because the recombination events took place in a resistant cell. Streptomycin-resistant donors and streptomycin-sensitive recipients, on the other hand, would not form recombinant colonies on medium containing streptomycin because the antibiotic would kill the cells where recombination events were occurring. Armed with this test to identify donor bacteria, Hayes discovered that they readily conferred donor ability on recipient cells and that the donor character spread through a recipient population much more quickly than individual cells could divide (17). He concluded that there must exist an infectious fertility factor, F, whose presence permitted F^+ donor *E. coli* cells to transfer DNA to F^- recipient cells and F must replicate independently of and more quickly than the bacterial chromosome. In this way, he defined the first bacterial plasmid and opened the door to a rigorous analysis of bacterial sexuality.

Hayes also found that donor cells only transferred a segment of their genomes to the recipient cells. Elie Wollman and François Jacob answered how the F plasmid mediated this partial transfer through their analysis of more potent donor strains. These were termed Hfr, for high frequency of recombination, because they could transfer certain markers orders of magnitude more frequently than F^+ donors. By examining a number of different Hfr strains and using a Waring blender to disrupt contacts between donor and recipient cells, they demonstrated that each Hfr strain transferred genetic markers in an oriented manner from a fixed point on the circular bacterial chromosome (18, 19). The transfer process took about 90 min to encompass the whole chromosome, and most mating pairs fell apart before this time, thus giving the partial transfer found by Hayes. The fixed point where transfer originated was interpreted as a site where the F plasmid inserted into the bacterial chromosome. Since different Hfr strains had different origins of transfer, insertion could occur at many locations. The genetic and physical connection between F and the chromosome in Hfr strains was established by the isolation of so-called F′ plasmids which replicated autonomously but carried segments of the bacterial chromosome that previously had been located adjacent to the Hfr origin of transfer (20). These F′ plasmids represented, in fact, some of the earliest examples of molecular cloning.

An even earlier example of molecular cloning came from studies of transduction by the temperate bacteriophage λ native to *E. coli* K-12. Transduction was first discovered in *Salmonella*, where it was found that some particles of phage P22 could transfer small fragments of the bacterial chromosome from one cell to another (21). P22 mediated a process called *generalized* transduction because it could transfer virtually any marker on the chromosome. When λ was tested for transduction, it behaved differently from P22: λ only transferred fragments carrying the *gal* (galactose utilization) marker, which had been located next to the λ prophage in Hfr crosses (22). λ transduction was called *specialized* transduction because it was limited to certain markers and because it displayed some peculiar features. Transductants received both phage and bacterial markers from the donor strain (23). These markers were incorporated into defective phage derivatives called λ*dgal*. When Campbell analysed the genetic structure of various λ*dgal* phages, he found that a particular region of phage DNA had always been substituted by a region of bacterial DNA (24), in a way analogous to contemporary cloning in λ vectors. To explain how these recombinant molecules originated, Campbell postulated that prophage insertion involved reciprocal recombination at specific sites between a circular phage molecule and the circular bacterial chromosome. Such an event would generate a larger circle with the (pro)phage DNA continuous with the chromosome at a specific position. Excision would occur by a complementary recombination event. Normally this excisive recombination would take place at the same sites as insertion, but occasionally excision would occur at other sites to yield λ*dgal* derivatives. These recombination steps comprised the well-known 'Campbell model' which explained specialized transduction, F insertion into the chromosome to form Hfrs, and aberrant F excision to generate F′ plasmids (25). Thus, even though they had many quite distinct properties, both F and λ were classified together as episomes because they displayed similar kinds of interactions with the bacterial chromosome (18). Further work on λ has greatly extended the Campbell model and identified special site-specific recombination enzymes which, together with their cognate specific DNA substrates, serve as the molecular tools for insertion and excision (26).

The DNA manipulation potential of F′ plasmids and specialized transducing phages was quickly appreciated among bacterial geneticists, and *in vivo* genetic engineering became a key tool in analysing genome structure and function starting in the middle to late 1960s (27). For example, temperature-sensitive replication mutants of F*lac* plasmids were isolated, and the ability of a temperature-sensitive F*lac* plasmid to rescue itself at the non-permissive temperature by insertion into the bacterial chromosome was exploited to transpose the *E. coli lac* operon to new genomic locations (28). Combined with specialized transduction of the transposed *lac* sequences (29), such methods permitted the isolation and purification of defined DNA sequences for molecular analysis even before restriction enzymes and all our other current genetic engineering tools were available (27, 30).

4. Transposable elements in bacteria

The parallels between bacterial episomes and maize controlling elements were only dimly appreciated in the late 1950s and early 1960s. The main reason that the relationship between them was not more widely recognized was the deep conviction that genetic rearrangements were aberrations. From this point of view, each example of genetic mobility was a peculiar exception to normal function, and the underlying molecular processes were stigmatized with the epithet of 'illegitimate recombination'. Nonetheless, the study of bacterial conjugation and transduction had firmly established in the minds of molecular biologists the concept of genetic elements which could migrate from one location to another through the genome. The use of phages and plasmids for *in vivo* genetic engineering experiments reinforced the appreciation of the capacity for genome plasticity. What finally made it generally evident that mobile genetic elements were an integral part of the genomic landscape was the discovery in bacteria of transposable elements whose only *raison d'être* seemed to be the ability to move DNA segments from one locus to another.

There are three strands to the prokaryotic transposable element story. They all began in the 1960s and came together with two key conferences in the mid-1970s. The first strand began with A. L. Taylor's discovery of a curious bacteriophage in Denver sewage. This phage was called Mu because it had the intriguing property of inserting its prophage at many different sites on the *E. coli* chromosome, sometimes causing mutations (31). Insertions had little sequence specificity. The mutations could occur in virtually any locus on the chromosome, and insertions into the *lac* operon were found at many different positions (31a, 32, 33). Since it could be propagated as a phage, Mu was particularly suitable for studying the molecular biology of this genetic promiscuity. Mu DNA could be extracted from purified phage particles, and phage genetics methods could be used to isolate mutants. The molecular studies showed that all aspects of Mu biology were intimately connected with its ability to move through the genome; recombination with different regions of the genome was essential to Mu's ability to replicate and lysogenize (34). Complementary genetic studies of how Mu interacted with different replicons in *E. coli* cells revealed it to be a powerful agent for rearranging the bacterial genome.

The second strand to the bacterial transposable element story was yet another example of multiple serendipity. Bacterial geneticists interested in the regulation of gene expression had isolated spontaneous pleiotropic mutations of the *lac* and *gal* operons (35–38). These mutations turned out to display unexpected properties (e.g. extreme polarity on the expression of cistrons downstream from the mutant site). They were not deletions because they could revert, but their responses to mutagens in reversion tests were quite different from those of well-known point mutations involving base substitutions or frame-shifts. Clearly, these spontaneous mutations represented a new kind of molecular event, and the hypothesis was proposed that they resulted from the insertion of additional DNA (38). Using specialized transducing phages to analyse the physical structure of the mutations, this hypothesis was quickly confirmed (39–42) and it was discovered that several distinct segments of DNA could each insert at a number of different sites. Accordingly, these segments were called insertion sequences, or IS elements (43, 44). The biological importance of IS elements became even more evident when they were found to be components of bacterial plasmids, such as F and drug-resistance determinants, and were located at key positions in those plasmids, including the sites where F recombines with the bacterial chromosome to form Hfrs (45, 46). It was becoming clear that IS elements played a major role in restructuring the bacterial genome through mutation and DNA transfer.

The third strand of the prokaryotic transposable element story involved the study of drug-resistance plasmids. The best-documented contemporary example we have of genetic change in evolution is the emergence of transmissible antibiotic resistance as the bacterial response to antibacterial chemotherapy. Following the discovery of R (resistance) plasmids in Japan in the later 1950s (47), there was intensive research on the genetics of antibiotic-resistance determinants. Many of these determinants displayed anomalous recombination behaviour and were ultimately found to transpose from one replicon to another (e.g. 48, 49). The term transposon was used to denote these transposable antibiotic-resistance determinants. It later became clear that many other kinds of phenotypic markers could also be carried on transposons, such as the ability to degrade specific substrates or the production of toxins and other virulence factors. Molecular analysis in the 1970s revealed that transposons came in (at least) two broad classes. One class, usually represented by the ampicillin transposon, Tn*3*, had a structure similar to that of IS elements but included one or more coding sequences for a specific phenotype inside the inverted repeats that marked the element's termini (48, 50). The other class consisted of compound elements, each of which contained a central coding sequence flanked by two copies of a particular IS element (49). Genetic analysis showed that the IS elements conferred genetic mobility on these compound elements and that any region of the bacterial genome could be incorporated into such a composite transposon. Because they played a key role in bacterial survival of the antibiotic onslaught, the biological utility of transposable elements could no longer be questioned.

5. Mobile elements in eukaryotes

The moment when it became clear that episomes and transposable elements were all related is well-defined: it was during a remarkable conference on bacterial plasmids held at Squaw Valley in 1975. At that meeting, held shortly after the famous Asilomar conference on the implications of recombinant DNA, there were reports on IS elements in plasmids, on the genetic activity of phage Mu, and on various antibiotic-resistance transposons. The idea of genetic engineering was very much in the air, and it was readily appreciated that bacteria were engineering their own genomes. The following year, stimulated directly by the Squaw Valley conference, the first meeting devoted specifically to DNA insertion elements, plasmids, and episomes was held at Cold Spring Harbor Laboratory (32). For the first time, mobile elements in bacteria and in higher organisms were discussed together from the perspective of mobility as a basic feature of all genomes. Genetic data on yeast, *Drosophila*, and animal viruses were invoked to show that the tremendous innate capacity for genome reorganization seen in maize and bacteria was a general phenomenon. In the years following the insertion element meeting, the application of recombinant DNA methods has abundantly verified McClintock's prediction that mutable loci and transposable elements would be found in the genomes of all organisms. The molecular work also revealed an unexpectedly wide variety of biochemical systems for restructuring genomes. The era of the Fluid Genome had arrived.

The chapters that follow will discuss specific examples of genome restructuring in some detail. Here it is perhaps useful to distinguish eukaryotic mobile genetic elements into two broad categories: DNA-based mobile elements and RNA-based ones (including retrotransposons).

5.1 DNA-based mobile elements

All of the mobile elements discussed above in bacteria and maize appear to operate exclusively at the DNA level. The biochemical activities responsible for mobility recognize structural features in the mobile elements, make cleavages both in those elements and at other genomic sites, and then religate the cleaved segments in new ways to create intact but rearranged DNA molecules. Examples of DNA-based elements appear to include the *P*, *hobo* and *FB* (foldback) elements in *Drosophila*, the Tc*1* element of *Caenorhabditis elegans*, and plant elements related to the *Ac* and *Spm* systems of maize (see relevant chapters in reference 51). It is important to note that not all DNA-based mobile elements fall into a single mechanistic class. We already know of several different biochemical mechanisms for the movement of DNA-based elements. In bacteria alone, for example, λ uses a reciprocal site-specific recombinase complex (26), Mu and Tn*3* have replicative transposition mechanisms (52, 53). Tn*7* has a non-replicative cut-and-paste transposition mechanism with a linear intermediate (54), and Tn*916* uses a non-replicative excision/re-insertion mechanism with a circular intermediate (55). In other words, genetic mobility operating at the DNA level appears to have arisen several times in evolution with distinct biochemistry.

5.2 Retrotransposons and other RNA-based mobile elements

In yeast, *Drosophila*, and vertebrates, the most abundant class of mobile elements appears to be those which move through RNA intermediates and utilize reverse transcription to insert at new genomic locations. A major subdivision of these retrotransposons are those which are structurally and mechanistically related to retroviruses, including the Ty elements of yeast, some of the most active *Drosophila* elements including *copia*, 412, B104, and *gypsy* (which are together responsible for the majority of spontaneous visible mutations; see reference 56), and elements in organisms as diverse as plants and slime moulds (see relevant chapters in reference 51). It is important to remember that retroviruses are themselves important mobile genetic elements, and that their pathogenic properties are often due to insertional mutagenesis or to transduction of modified cellular sequences in the form of viral oncogenes (57). Interestingly, the chromosomal integration of reverse-transcribed retroviral DNA is mechanistically similar to steps in bacteriophage Mu transposition (53, 58). As with the DNA-based elements, there is more than one pathway for RNA-based mobility, and another major group of retrotransposons lack the characteristic terminal repeats of retroviruses. These include the LINEs (long interspersed nucleotide elements; see reference 59), which encode their own reverse transcriptase, and the SINEs (short interspersed nucleotide elements; see reference 60) which do not encode reverse transcriptase). In addition, as discussed in Chapter 7, a new class of reverse-transcribed elements, called retrons, has also recently been discovered in bacteria.

6. Genome reorganization as a biological process

Many mobile genetic elements were initially detected by recombinant DNA analysis as repetitive elements. Molecular studies also revealed another aspect of genomic fluidity in the form of DNA rearrangements that accompany cellular differentiation. These developmental rearrangements involve a wide variety of biochemical mechanisms. For example, both *Salmonella* and *Neisseria* undergo phase variations (on–off switches) affecting the biogenesis of surface structures, but the *Salmonella* system uses a site-specific recombinase (61) while the *Neisseria* system uses homology-dependent recombination (62). In the vertebrate immune system, at least four mechanistically different events are involved in assembling and modifying the sequences which encode immunoglobulins during B lymphocyte maturation (63, 64). Some organisms have very regular cycles of genome reorganization during somatic development or gametogenesis. Many of these cycles are lumped together under the rubric of 'chromatin diminution', but that designation includes many mechanistically different events. In some cases, pieces of chromosomes are excised during development (e.g. 65), while in other cases whole chromosomes are discarded (e.g. 66). Perhaps the most spectacular example of chromatin diminution occurs during macronuclear development in ciliated protozoa (67). In this remarkable process, the entire germline genome is fragmented into smaller pieces of DNA; some of these are then joined together in new arrangements, and the thousands of resulting DNA segments are each capped with telomeres to produce mini-chromosomes that direct the synthesis of RNA (68).

What our rapidly expanding knowledge of mobile genetic elements and developmental DNA rearrangements is telling us is that cells have at their disposal a large tool-box of biochemical activities for restructuring their genomes. In this context, it is relevant to bear in mind that almost all of the biochemical tools used in human genetic engineering are extracted from one or another cell type. Most of these cellular DNA rearrangement activities are quite sophisticated and involve multiple protein and nucleic acid components organized into intricate dynamic three-dimensional complexes, such as the λ intasome (26, 69) or the retroviral particle (57, 70). Like all biochemical systems, those carrying out DNA rearrangements are subject to cellular and organismal regulatory networks. The ability of these networks to modulate genetic changes in response to developmental or environmental cues is reflected in specific cases, such as the developmental specificity of immune system rearrangements (64) or the starvation-induced activation of a Mu prophage to produce coding sequence fusions that permit the utilization of an alternative growth substrate (71–73). In other words, the discovery of multiple sophisticated biochemical systems for DNA reorganization means that genetic change is a normal function of living cells (74). There is nothing 'illegitimate' about the many nucleases, recombinases, ligases, and other rearrangement activities we are uncovering, and their adaptive functions can be studied by the same genetic and biochemical approaches used to elucidate other aspects of cell and organismal biology.

7. Integral view of genome function and evolution

McClintock was quite prophetic when she asserted that mobile genetic elements would profoundly alter our thinking about genome organization and evolution. Not only has the Fluid Genome model replaced the Constant Genome one, but our thinking about genome structure and function has also undergone a sea-change. McClintock's insight that genetic loci contained distinct, variable regulatory regions had been borne out by numerous studies of the molecular basis of gene expression, especially work on transcriptional regulation and the processing of RNA. Our current conception of the typical genetic locus as a mosaic of exons, introns, and 5′ and 3′ regulatory regions, each composed of its own array of domains and binding sites, is totally different from the classic theory of the unitary gene formed in the pre-DNA era. We now realize that repetitive sequence elements, such as binding sites for transcription factors, are present at multiple unlinked loci and serve as one of the physical bases for co-ordinately regulated multigenic networks. We also understand that other repetitive elements, such as centromeres and telomeres, play key roles in maintaining the physical organization of the genome.

Comparing this view of the genome as an interactive network of multiple sequence elements with the classical idea of the genome as composed of autonomous genetic units linked together like beads on a string is like comparing our current concept of atomic structure with that of the pre-quantum mechanics era. As we near the end of the twentieth century, molecular genetics has led us into a different conceptual universe. It would be strange if such a major change in our thinking about basic genetic organization did not also influence our understanding of evolutionary processes. In order for the integrated mosaic genome to make evolutionary sense, there must exist mechanisms for large-scale, rapid reorganization of diverse sequence elements into new configurations. By analogy with computer-based systems, evolution could be envisaged as involving changes from one system architecture to another (75). When we thought of genetic change only in terms of point mutations and random events, massive genome reorganization was inconceivable, but the study of mobile elements has now made the inconceivable very real. McClintock (76) observed major genome restructuring in some of the plants emerging from the breakage–fusion–bridge cycle, and we now have many other examples, such as ciliate macronuclear development and multiple pre-meiotic germline transpositions in *Drosophila* hybrid dysgenesis (77). As we incorporate our knowledge of DNA rearrangements into basic theories of heredity, it may be useful to think of evolution as a natural genetic engineering process (78). If this viewpoint proves itself intellectually robust enough to direct an effective research agenda, then mobile genetic elements will have moved from the fringes to the centre of modern biological thought.

8. Barbara McClintock, 1902–1992

This introduction was drafted shortly before the death of Barbara McClintock on 2 September 1992, at the age of 90. Since McClintock was the lone pioneer in the study of mobile genetic elements for so many years and since she has had such a profound impact on modern genetics, it is fitting to add a few words of tribute to her in this introduction. McClintock was one of the outstanding figures in modern science. From the establishment of chromosomes as the physical carriers of hereditary determinants to the analysis of nuclear networks governing gene expression, her 69 year career was an integral part of the genetic revolution which is still transforming our understanding of life.

We are not yet in a position to evaluate the full significance of Barbara McClintock's scientific accomplishments. We recognize that her work on transposable elements revolutionized our thinking about genome stability and genome reorganization. But the implications of her observations that cells can rapidly detect the presence of broken chromosomes and efficiently repair the breaks remain to be fully explored. Likewise, genetic theory has not yet fully incorporated her discovery of repetitive mobile genetic systems that can alter the developmental expression of any genetic locus and can create control networks involving unlinked loci.

The main reasons that McClintock's insights are still outside the mainstream have more to do with attitudes than with data. Even though her thinking was far more sophisticated, a common misconception has persisted that McClintock thought of controlling element insertions and excisions as the chief mechanism of developmental gene regulation. While it is true that excisions did occur in a regulated manner and thus gave rise to developmental patterns, she also documented many novel patterns of gene expression that did not involve mutational events. For McClintock, the main point was that each new controlling element insertion (or change in the structure of a resident element) modified a particular genetic locus and brought it under the control of a wide repertoire of regulatory mechanisms. Similar modifications at two or more loci would create a genetic network. Another obstacle to broader acceptance of McClintock's perspective is the widespread tendency to explain hereditary phenomena in terms of independent genetic units, a holdover from the days when Mendelian segregations were the basic methods of analysis. In contrast, McClintock thought of the genome as a complex unified system exquisitely integrated into the cell and the organism, and her work with controlling elements revealed some of the physical mechanisms that were at the basis of genomic integration.

McClintock's vision extended beyond the genome. One of her most challenging ideas is the concept of 'smart cells' (something she used to bring up gingerly in lectures at the end of her career). Behind this concept lay decades of experience. Her own work involved tracing the development of tens of thousands of maize plants in intimate detail, and she was well aware of the work of other scientists, from the nineteenth century pioneers all the way up to contemporary molecular cell biologists. She was deeply impressed by the ability of cells to sense internal and external cues, evaluate them, and respond with actions appropriate for survival and morphogenesis. How this monitoring and decision-making operate was, she felt, a key area for future exploration.

What made McClintock so special? The answer lies in her complete intellectual freedom. The courage to say, 'I do not understand', and the courage to investigate the unexplainable were at the heart of her remarkable success. Many scientists have been upset because Barbara McClintock characterized herself as a mystic. But to her, mystic did not mean someone who mystifies. Instead, for Barbara McClintock, a mystic was someone with a deep awareness of the mysteries posed by natural phenomena. Mystification came, in her view, when we tried to use our current concepts to explain phenomena that demanded new ways of thinking.

Barbara McClintock occupies a unique place in the history of biology. Her work spanned almost the entire twentieth century. A main participant in many aspects of this century's revolutionary exploration into the physical basis of heredity, she began her studies only two decades after the rediscovery of Mendelism. Her observations on genetic networks and genome reorganization have defined problems to be addressed in the twenty-first century. It is possible that McClintock will one day be seen as the key figure marking the transition from the naturalistic biology of the nineteenth century to the informational biology of the future. Her rich scientific legacy will reward continued study for decades to come.

References

1. Kuhn, T. (1960) *The Structure of Scientific Revolutions*. University of Chicago Press, Chicago.
2. Shapiro, J. A. (ed.) (1983) *Mobile Genetic Elements*. Academic Press, New York.
3. McClintock, B. (1987) *The Discovery and Characterization of Transposable Elements: the Collected Papers of Barbara McClintock*. Garland, New York.
4. McClintock, B. (1938) The production of homozygous deficient tissues with mutant characteristics by means of the aberrant mitotic behavior of ring-shaped chromosomes. *Genetics*, **21**, 315.
5. McClintock, B. (1956) Controlling elements and the gene. *Cold Spring Harbor Symp. Quant. Biol.*, **21**, 197.
6. McClintock, B. (1951) Chromosome organization and genetic expression. *Cold Spring Harbor Symp. Quant. Biol.*, **16**, 13.

7. McClintock, B. (1961) Some parallels between gene control systems in maize and in bacteria. *Am. Nat.*, **95,** 265.
8. McClintock, B. (1950) The origin and behavior of mutable loci in maize. *Proc. Natl Acad. Sci. USA*, **36,** 344.
9. McClintock, B. (1953) Induction of instability at selected loci in maize. *Genetics*, **38,** 579.
10. McClintock, B. (1965) The control of gene action in maize. *Brookhaven Symp. Biol.*, **18,** 162.
11. McClintock, B. (1978) Development of the maize endosperm as revealed by clones. *Symp. Soc. Dev. Biol.*, **36,** 217.
12. Shapiro, J. A. (1992) Kernels and colonies: the challenge of pattern. In *The Dynamic Genome*. N. Federoff and D. Botstein (eds). Cold Spring Harbor Laboratory Press, Cold Spring Harbor, NY, p. 213.
13. Federoff, N. (1989) Maize transposable elements. In *Mobile DNA*. D. E. Berg and M. M. Howe (eds). American Society for Microbiology, Washington, DC, p. 375.
14. McClintock, B. (1956) Intranuclear systems controlling gene action and mutation. *Brookhaven Symp. Biol.*, **8,** 58–74.
15. Lederberg, J. and Tatum, E. L. (1946) Novel genotypes in mixed cultures of biochemical mutants of bacteria. *Cold Spring Harbor Symp. Quant. Biol.*, **1,** 113.
16. Hayes, W. (1952) Recombination in *Bact. coli* K-12: unidirectional transfer of genetic material. *Nature*, **169,** 118.
17. Hayes, W. (1953) Observations on a transmissible agent determining sexual differentiation in *Bact. coli*. *J. Gen. Microbiol.*, **8,** 72.
18. Jacob, F. and Wollman, E. (1961) *Genetics and the Sexuality of Bacteria*. Academic Press, New York.
19. Wollman, E., Jacob, F., and Hayes, W. (1956) Conjugation and genetic recombination in *Escherichia coli*. *Cold Spring Harbor Symp. Quant. Biol.*, **21,** 141.
20. Adelberg, E. A. and Burns, S. N. (1960) Genetic variation in the sex factor of *Escherichia coli*. *J. Bacteriol.*, **79,** 321.
21. Zinder, N. D. (1953) Infective heredity in bacteria. *Cold Spring Harbor Symp. Quant. Biol.*, **18,** 261.
22. Morse, M. L. (1954) Transduction of certain loci in *Escherichia coli* K-12. *Genetics*, **39,** 984.
23. Morse, M. L., Lederberg, E. M., and Lederberg, J. (1956) Transductional heterogenotes in *Escherichia coli*. *Genetics*, **41,** 758.
24. Campbell, A. (1959) Ordering of genetic sites in bacteriophage λ by the use of galactose-transducing defective phages. *Virology*, **9,** 293.
25. Campbell, A. (1962) Episomes. *Adv. Genet.*, **11,** 101.
26. Landy, A. (1990) Dynamic, structural and regulatory aspects of λ site-specific recombination. *Annu. Rev. Biochem.*, **58,** 913.
27. Shapiro, J. A. (1982) Mobile genetic elements and reorganization of prokaryotic genomes. In *Genetics of Industrial Microorganisms*. Y. Ikeda, and T. Beppu (eds). Kodansha, Tokyo, p. 9.
28. Jacob, F., Brenner, S., and Cuzin, F. (1963) On the regulation of DNA replication in bacteria. *Cold Spring Harbor Symp. Quant, Biol.*, **28,** 329.
29. Beckwith, J. R., Signer, E. R., and Epstein, W. (1966) Transposition of the *lac* region of *E. coli*. *Cold Spring Harbor Symp. Quant. Biol.*, **31,** 393.
30. Shapiro, J. A., MacHattie, L., Eron, L., Ihler, G., Ippen, K., Beckwith, J. R., Arditti, R., Reznikoff, W., and MacGillivray, R. (1969) The isolation of pure *lac* operon DNA. *Nature*, **224,** 768.

31. Taylor, A. L. (1963) Bacteriophage-induced mutation in *E. coli*. *Proc. Natl Acad. Sci. USA*, **50,** 1043.
31a. Bukhari, A. I. and Zipser, D. (1972) Random insertion of Mu-1 DNA within a single gene. *Nature New Biol.*, **236**, 240.
32. Bukhari, A. I., Shapiro, J. A., and Adhya, S. L. (eds) (1977) *DNA Insertion elements, Plasmids and Episomes*. Cold Spring Harbor Laboratory Press, Cold Spring Harbor, NY.
33. Daniell, E., Roberts, R., and Abelson, J. (1972) Mutations in the lactose operon caused by bacteriophage Mu. *J. Mol. Biol.*, **69,** 1.
34. Toussaint, A. and Resibois, A. (1983) Phage Mu: transposition as a life-style. In *Mobile Genetic Elements*. J. A. Shapiro (ed.). Academic Press, New York, p. 105.
35. Adhya, S. and Shapiro, J. A. (1969) The galactose operon of *E. coli* K-12. I. Structural and pleiotropic mutations of the operon. *Genetics*, **62,** 231.
36. Malamy, M. H. (1967) Frameshift mutations in the lactose operon of *E. coli*. *Cold Spring Harbor Symp. Quant. Biol.*, **31,** 189.
37. Jordan, E., Saedler, H., and Starlinger, P. (1967) Strong polar mutations in the transferase gene of the galactose operon of *E. coli*. *Mol. Gen. Genet.*, **100,** 296.
38. Shapiro, J. A. (1967) The structure of the galactose operon in *Escherichia coli* K-12. Doctoral dissertation, Cambridge University.
39. Shapiro, J. A. (1969) Mutations caused by the insertion of genetic material into the galactose operon of *Escherichia coli*. *J. Mol. Biol.*, **40,** 93–105.
40. Jordan, E., Saedler, H., and Starlinger, P. (1968) O^0 and strong polar mutations in the *gal* operon are insertions. *Mol. Gen. Genet.*, **102,** 353.
41. Malamy, M. H., Fiandt, M., and Szybalski, W. (1972) Electron microscopy of polar insertions in the *lac* operon of *Escherichia coli*. *Mol. Gen. Genet.*, **119,** 207.
42. Hirsch, H. J., Saedler, H., and Starlinger, P. (1972) Insertion mutations in the control region of the galactoase operon of *E. coli*. II. Physical characterization of the mutations. *Mol. Gen. Genet.*, **115,** 266.
43. Fiandt, M., Szybalski, W., and Malamy, M. H. (1972) Polar mutations in *lac, gal* and phage λ consist of a few DNA sequences inserted with either orientation. *Mol. Gen. Genet.*, **119,** 223.
44. Hirsch, H. J., Starlinger, P., and Brachet, P. (1972) Two kinds of insertions in bacterial genes. *Mol. Gen. Genet.*, **119,** 191.
45. Davidson, N., Deonier, R. C., Hu, S., and Ohstubo, E. (1975) Electron microscope heteroduplex studies of sequence relations among plasmids of *Eschericha coli*. X. Deoxyribonucleic acid sequence organization of F and of F-primes and the sequences involved in Hfr formation. In *Microbiology 1974*. D. Schlessinger (ed.). American Society for Microbiology, Washington, DC, p. 56.
46. Hu, S., Ohtsubo, E., Davidson, N., and Saedler, H. (1975) Electron microscope heteroduplex studies of sequence relations among bacterial plasmids: identification and mapping of the insertion sequences IS1 and IS2 in F and R plasmids. *J. Bacteriol.*, **122,** 764.
47. Watanabe, T. (1963) Infectious heredity of multiple drug resistance in bacteria. *Bacteriol. Rev.*, **27,** 87.
48. Heffron, F. (1983) Tn*3* and its relatives. In *Mobile Genetic Elements*. J. A. Shapiro (ed.). Academic Press, New York, p. 223.
49. Kleckner, N. (1983) Transposon Tn*10*. In *Mobile Genetic Elements*. J. A. Shapiro (ed.). Academic Press, New York, p. 261.
50. Sherratt, D. (1989) Tn*3* and related transposable elements: site-specific recombination

and transposition. In *Mobile DNA*. D. E. Berg and M. M. Howe (eds). American Society for Microbiology, Washington, DC, p. 163.
51. Berg, D. E. and Howe, M. M. (eds) (1989). *Mobile DNA*. American Society for Microbiology, Washington, DC.
52. Arthur, A. and Sherratt, D. (1979) Dissection of the transposition process: a transposon-encoded site-specific recombination system. *Mol. Gen. Genet.*, **175,** 267.
53. Shapiro, J. (1979) A molecular model for the transposition and replication of bacteriophage Mu and other transposable elements. *Proc. Natl Acad. Sci. USA*, **76,** 1933.
54. Bainton, R., Gamas, P., and Craig, N. (1991) Tn*7* transposition *in vitro* proceeds through an excised transposon intermediate generated by staggered breaks in DNA. *Cell*, **65,** 805.
55. Caparon, M. G. and Scott, J. R. (1989) Excision and insertion of the conjugative transposon Tn*916* involves a novel recombination mechanism. *Cell*, **59,** 1027.
56. Green, M. M. (1988) Mobile DNA elements and spontaneous gene mutation. In *Eukaryotic transposable elements as mutagenic agents*, E. Lambert, J. F. McDonald, I. B. Weinstein (eds), Banbury Report Vol. 30, Cold Spr. Harb. Laboratory, p. 41.
57. Varmus, H. (1983) Retroviruses. In *Mobile Genetic Elements*. J. A. Shapiro (ed.). Academic Press, New York, p. 411.
58. Fujiwara, T. and Mizuuchi, K (1988) Retroviral DNA integration: structure of an integration intermediate. *Cell*, **54,** 497.
59. Hutchison, C. A., Hardies, S. C., Loeb, D. D., Shehee, W. R., and Edgell, M. H. (1989) LINES and related retroposons: long interspersed repeated sequences in the eucaryotic genome. In *Mobile DNA*. D. E. Berg and M. M. Howe (eds). American Society for Microbiology, Washington, DC, p. 593.
60. Deininger, P. L. (1989) SINES: short interspersed repeated DNA elements in higher eucaryotes. In *Mobile DNA*. D. E. Berg and M. M. Howe (eds). American Society for Microbiology, Washington, DC, p. 619.
61. Glasgow, A. C., Hughes, K. T., and Simon, M. I. (1989) Bacterial DNA inversion systems. In *Mobile DNA*. D. E. Berg and M. M. Howe (eds). American Society for Microbiology, Washington, DC, p. 637.
62. Swanson, J. and Koomey, J. M. (1989) Mechanisms for variation of pili and outer membrane protein II in *Neisseria gonorrhoeae*. In *Mobile DNA*. D. E. Berg and M. M. Howe (eds). American Society for Microbiology, Washington, DC, p. 743.
63. Alt, F. W., Blackwell, T. K., and Yancopoulos, G. D. (1987) Development of the primary antibody repertoire. *Science*, **238,** 1079.
64. Blackwell, T. K. and Alt, F. W. (1989) Mechanism and developmental program of immunoglobulin gene rearrangement in mammals. *Annu. Rev. Genet.*, **23,** 605.
65. Beerman, S. (1977) The diminution of heterochromatic chromosomal segments in *Cyclops* (Crustacea, Copepoda). *Chromosoma*, **60,** 297.
66. Crouse, H. V. (1960) The controlling element in sex chromosome behavior in *Sciara*. *Genetics*, **45,** 519.
67. Gall, J. (1986) *The Molecular Biology of Ciliated Protozoa*. Academic Press, Orlando, FL.
68. Yao, M.-C. (1989) Site-specific chromosome breakage and DNA deletion in ciliates. In *Mobile DNA*. D. E. Berg and M. M. Howe (eds). American Society for Microbiology, Washington, DC, p. 715.
69. Echols, H. (1986) Multiple DNA–protein interactions governing high-precision DNA transactions. *Science*, **233,** 1050.

70. Varmus, H. and Brown, P. (1989) Retroviruses. In *Mobile DNA*. D. E. Berg and M. M. Howe (eds). American Society for Microbiology, Washington, DC, p. 53.
71. Shapiro, J. A. (1984) Observations on the formation of clones containing *araB–lacZ* cistron fusions. *Mol. Gen. Genet.*, **194,** 79.
72. Cairns, J., Overbaugh, J., and Miller, S. (1988) The origin of mutants. *Nature,* **335,** 142.
73. Shapiro, J. A. and Leach, D. (1990) Action of a transposable element in coding sequence fusions. *Genetics,* **126,** 293.
74. Shapiro, J. A. (1985) Mechanisms of DNA reorganization in bacteria. *Int. Rev. Cytol.*, **93,** 25.
75. Shapiro, J. A. (1991) Genomes as smart systems. *Genetica,* **84,** 3.
76. McClintock, B. (1978) Mechanisms that rapidly reorganize the genome. Stadler Genetics Symp., **10,** 25.
77. Woodruff, R. C. and Thompson, J. N. (1992) Have premeiotic clusters of mutation been overlooked in evolutionary theory? *J. Evol. Biol.*, **5,** 457.
78. Shapiro, J. A. (1992) Natural genetic engineering in evolution. *Genetica,* **86,** 99.

2 Mechanisms of DNA transposition

RONALD H. A. PLASTERK

1. Introduction

All organisms studied in any detail have been found to contain transposons, discrete segments of DNA that can move within genomes. Transposons can be viewed as molecular parasites, the smallest units of selection: single (or a few) genes that move around by using all functions of the host (building blocks, energy, the replication apparatus), and that only provide the factors that are needed for the recognition of the border between themselves and their host DNA, and for the initiation of events that result in their own spread within the host genome. The only necessary explanation for their existence is that they have survived, but this does not exclude the possibility that, like many other parasites, transposons could be anywhere on the scale between parasitism and symbiosis, and thus be for their hosts either a nuisance, or an indispensable helper, or both.

The notion of transposons as molecular parasites has been met with scepticism, most notably from their discoverer Barbara McClintock, who referred to them more fondly as 'controlling elements', and saw them playing an essential role in organismal development (1). One can sympathize with criticism of those (rare) molecular biologists who think we will understand an organism when we have sequenced it, and who refer to anything they do not understand as 'junk', including all repetitive, non-coding, and transposable DNA. However, it is not immediately clear what is so offensive about the idea that transposons are complete units of genetic selection, with a lifestyle intimately interwoven with their cellular environment, with no other purpose in life than to survive. Transposons have the same reasons for existence as humans: we are not here because we are nice, useful, or good, but because we have survived.

What constraints does the parasite lifestyle put on the mechanism of transposition?

1. The transposon has to be able to spread in a replicative fashion. One possibility is that the transposition reaction involves a replication step (as for phage Mu). However, the actual jump does not need to be replicative; it is sufficient that the overall frequency of jumping is higher from replicated to non-replicated

DNA, because in that case a non-replicative jump will, as a result of a free ride on the cellular replication machine, result in a net gain of one transposon per two genomes (see Fig. 1a). A free ride also results when a non-replicative jump is followed by re-insertion of a *new* copy of the element into the *old* ('donor') site as a result of double-strand break repair (Fig. 1b; also see below).

2. The transposon should carry the steps that initiate transposition at least to the point where the cell can do nothing but finish the process. This implies that all transposon-specific steps have to be mediated by transposon encoded factors. If you are selfish, you have to take care of yourself, because no-one else has a reason to do it. In practice it means that all transposons encode at least one factor that in one way or another is involved in mediating the jump, and could therefore be referred to as the transposase. Note that among transposons there is the ultimate parasitism of non-autonomous transposons, usually deletion derivatives, that are *cis*-proficient, but *trans*-deficient, and depend on transposase provided by their autonomous relatives.

3. The transposon should distinguish between the 'mine' and the 'thine' as precisely as possible. Constant loss of sequence from the end of the element with every jump, or continuous growth, would mean a quick end to the existence of the element. In practice, the boundaries between transposon and flanking DNA are usually precise, and only for some retro-elements without long terminal repeats (LTRs), (such as LINES) is some end variation observed.

4. Transposition should be regulated, since too high a frequency of transposition will kill the host, and thus the transposons in it. Regulation is exerted at many levels: transcription and translation of transposase, modification of transposon DNA ends, competition of repressor proteins for binding to transposon ends, etc.

In this review, I will give an overview of the mechanisms of transposition, and will order my remarks by theme, rather than by organism. An excellent catalogue of transposons, ordered by host organisms and by transposon type, exists (2); it is not nearly outdated, and the reader is referred to this, and references therein, for more detailed discussions (other recent reviews are, among others, references 3–7). A choice of topics is dependent on the availability of experimental support for mechanistic models, and by the author's ability to understand these models, which implies a strong bias towards simplicity. There will also be an inevitable idiosyncrasy in the choice of illustrations of general principles, for which I apologize beforehand.

2. The transposition reaction

Transposition involves the breakage and formation of phosphodiester bonds. This section will discuss the mechanisms of breakage and reunion, in light of the

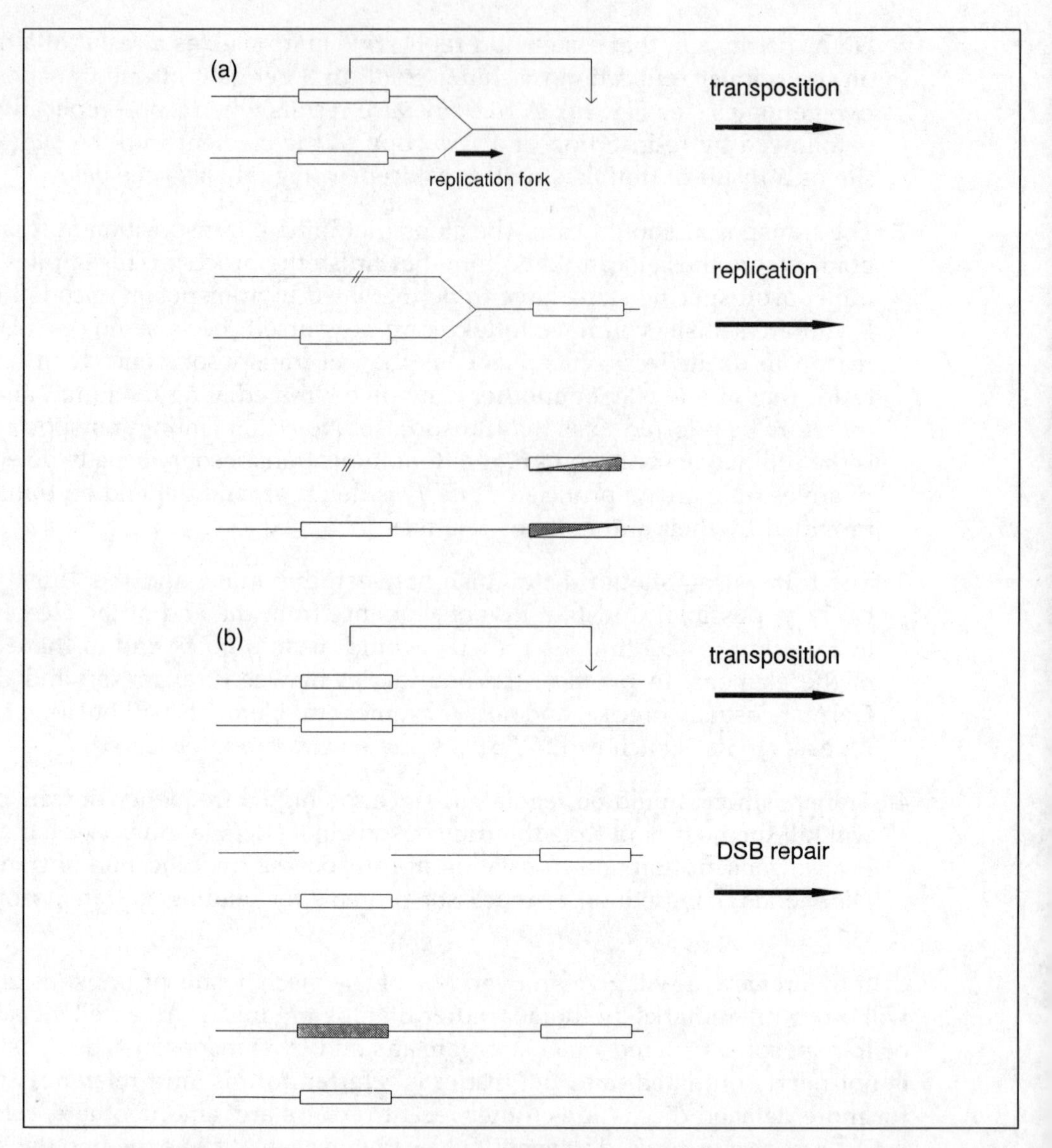

Fig. 1 Transposon spread through non-replicative transposition. (a) Transposition from replicated to not yet replicated DNA results in a net gain of one transposon for every two haploid genomes. Note that although the transposition is conservative, after replication two semi-conservatively replicated elements are at the new integration site, and one unaltered element and one empty site are left in the donor region. (b) Transposition followed by DNA double strand break (DSB) repair results in a net gain of one element for every two haploid genomes. Note that an unreplicated element is found at the new insertion site, and—depending on the precise mechanism of DSB repair—one unaltered and one completely new element are in the donor region.

surprising finding that, thus far, the chemistry of this reaction is identical for replicatively jumping bacterial transposon Mu, non-replicatively jumping bacterial transposons Tn*10* and Tn*7*, the retrotransposons (such as HIV), and probably the *Drosophila* transposon *P*. All of these reactions have been studied *in vitro*. The paradigm is the Mu transposon, whose transposition is best understood at present, primarily owing to the work of Mizuuchi and co-workers at the NIH, Bethesda (for review see reference 6).

2.1 Donor cut

In all cases the chemistry of the donor cut involves the oxygen atom at the 3′ end of the transposon DNA. The step that initiates transposition is the release of the 3′-OH group, at both ends of the transposon, from the phosphate in the flanking DNA (the 'donor' cut). For Mu and HIV retrovirus this donor cut involves single strand DNA cuts, as a result of the hydrolysis of one phosphodiester bond at the 3′ end of both transposon strands. Tn*10*, and probably the *Drosophila* transposon *P*, nematode transposon Tc*1*, and possibly many other elements, hydrolyse bonds in both strands at each end, thus releasing the transposon completely from the flanking DNA, but Mu remains bound to the flanks by the bonds at the 5′ ends (see Fig. 2).

2.1.1 Retrovirus integration

This is not the place to describe the retroviral replication cycle in detail (a detailed description is provided in reference 8, a specific description for yeast Ty elements in reference 9, and for non-LTR retro-elements in reference 10), but a brief intermezzo is necessary to summarize the steps that precede the donor cut: an integrated retrovirus genome is transcribed into genomic RNA, which can subsequently be incorporated in viral particles. The retrovirus has to reconcile the above-mentioned requirement of precise distinction between itself and flanking DNA, with the potentially slippery nature of transcription initiation and termination. Precise termination of transcription at the transposon–flank boundary, irrespective of the sequence of the flank, is probably too much to be asked, and the element has solved this by initiating and terminating transcription safely within its own territory, far away from the boundaries with flanking DNA. As a result, it has to replace the lost sequences in a later stage. These sequences are part of the LTRs, present at both transposon ends, and thus can be copied from the 3′ end of the genomic RNA to the 5′ end, and vice versa. This reaction is carried out by the enzyme reverse transcriptase. It involves a few template jumps, which are clearly explained by Varmus and Brown (ref. 8). The product resulting from this cDNA synthesis is a linear double-stranded DNA that is almost identical to the integrated provirus, except for a few (usually two) additional base-pairs at both ends. At this point in the retroviral replication cycle we can go back to transposition.

The donor cut chemistry is the same for phage Mu and retro-elements. A nick is made at the junction between the 3′-terminal nucleotide of the transposon and

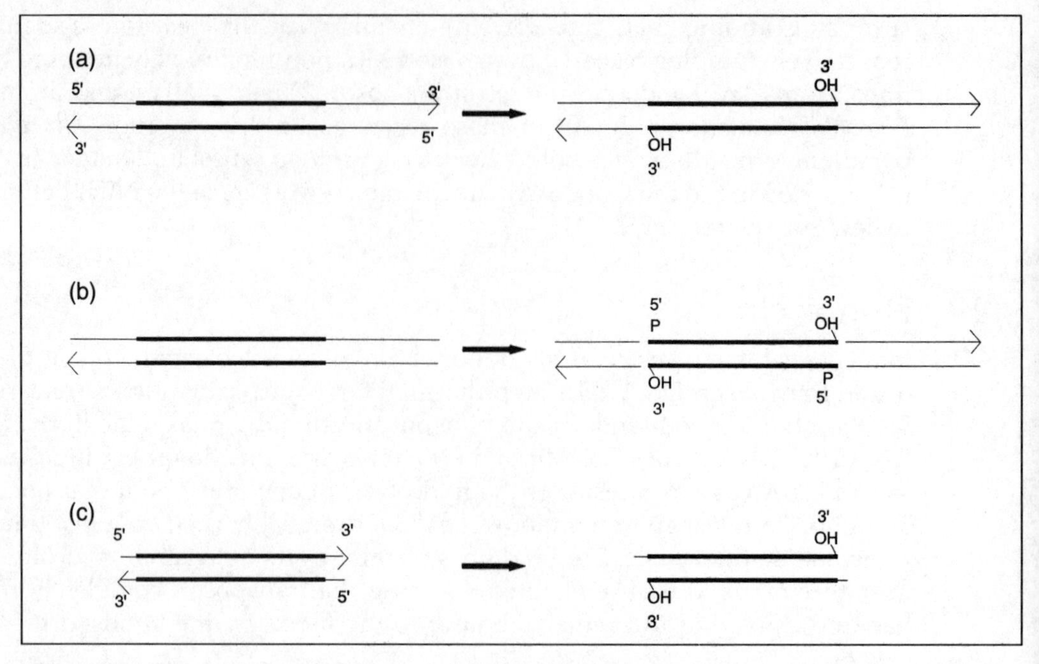

Fig. 2 The donor cut reaction. (a) Cutting of 3′ ends only (see text for explanation). (b) Cutting of both ends. Shown here is the situation in which cutting leaves blunt ends (see text). (c) Retrovirus donor cut. Note that, except for the length of the flank, the reaction is identical to that in panel (a). Note that whereas in Figs 1 and 3a a single line represents a double-stranded DNA molecule, it here represents a single DNA strand. The polarity of the strands is indicated: arrow heads are 3′ ends.

the two flanking nucleotides or flanking DNA, by hydrolysis of a phosphodiester bond, leaving a 3′-OH on the viral DNA and a 5′ phosphate on the released two nucleotides. It has been shown that the HIV transposase (integrase, or IN) mediates hydrolysis by making the bond sensitive to nucleophilic attack. The nucleophile is usually water, but can also be one of several alcohols, and alcoholic amino acids (11), or even the hydroxyl group at the 3′ end of the two nucleotides that are removed (12). The stereochemical course of this last reaction, which forms a circular dinucleotide, is in agreement with a single nucleophilic attack by the 3′-OH on the phosphodiester bond (12).

Nothing is known about the chemistry of the cut at the 5′ ends of the DNA of those transposons that are completely excised from the host genome (such as Tn*10* in bacteria, *P* in *Drosophila*, Tc*1* in nematodes). However, since this reaction is presumably also carried out by the transposase, it probably involves a similar chemistry. The simplest model would involve a dimer of transposase binding a transposon end: one monomer in each strand. One monomer binds inside the transposon, and the other binds to the flank, such that they are symmetrically positioned around the junction between transposon and flank. Recognition of the transposon end DNA by one monomer may activate both monomers, and trigger cutting of both strands (13). Whereas Tn*10* excises as a blunt-ended DNA as a result of cutting at identical positions in opposite strands, there is some evidence that Tec elements in ciliated protozoa may excise by a staggered cut, leaving single-stranded overhangs (14, 15). This has no relevance for further steps, as long as the cuts at the 3′ ends are made precisely at the transposon boundaries, assuming that (as discussed below, see Section 2.3) 3′-OH groups are central in the transposition process.

2.2 What happens to the flank?

2.2.1 After a double-strand cut at both transposon ends

As described, the donor cut either releases both strands of the transposon from its flank, or releases only the 3′ ends. In the first case, the donor site is empty, left with a double-strand DNA break. There is no reason to think that it is really a concern of the transposon what happens to the break; the transposon can excise and go to a new neighbourhood, and meanwhile the cell will deal with the break. Several possibilities exist, perhaps even within one organism:

1. If nothing happens to repair the break, the replicon will, in all probability, be lost. For plasmids in bacteria, this is not inconceivable, but in eukaryotes this would normally result in the death of the cell.

2. It has been shown that the excision of *P* elements in the germline of *Drosophila*, and of Tc*1* elements in *Caenorhabditis elegans*, is followed by template-dependent double-strand break (DSB) repair (16–19). Apparently, here the cell behaves like a bureaucrat, who does not want to act without proper instruction in fear of making a mistake, and therefore searches for a template. Usually, the instructing template is either the sister chromatid or the homologous chromosome. If the template is the homologous chromosome, and if that does not contain a transposon at that position (in other words if the transposon allele was heterozygous) the DSB repair results in precise loss of the element. If on the other hand the transposon allele is homozygous or if the sister chromatid is used as template, then DSB repair will result in the re-introduction of a *new* copy of the transposon into the *old* site (see Fig. 1b). Loss of the element is, therefore, even after precise excision, a rare event. If it occurs at all in homozygotes, it will be the result of incorrect DSB repair, by which the element is rarely precisely lost, and is usually accompanied by re-introduction of some sequences of one or both transposon ends (the so-called footprint, see Fig. 3a). These footprints have also been observed for plant transposons (20), suggesting that a similar repair mechanism can play a role here. It should be noted that this model explains how *precise* excision of a transposon results in *imprecise* loss of the element. This had not been envisaged before the first publication of Engels and co-workers (ref. 16), and unfortunately is still not fully realized in recent literature on transposition. Of course, the model that has now been proven for *P* elements in *Drosophila* (16, 18) and Tc*1* elements in *C. elegans* (17, 19) does not need to apply to other transposons that leave footprints, but the point is that imprecise transposon loss is not necessarily the same as imprecise excision. Since the excision reaction has, in all genetic and biochemical experiments,

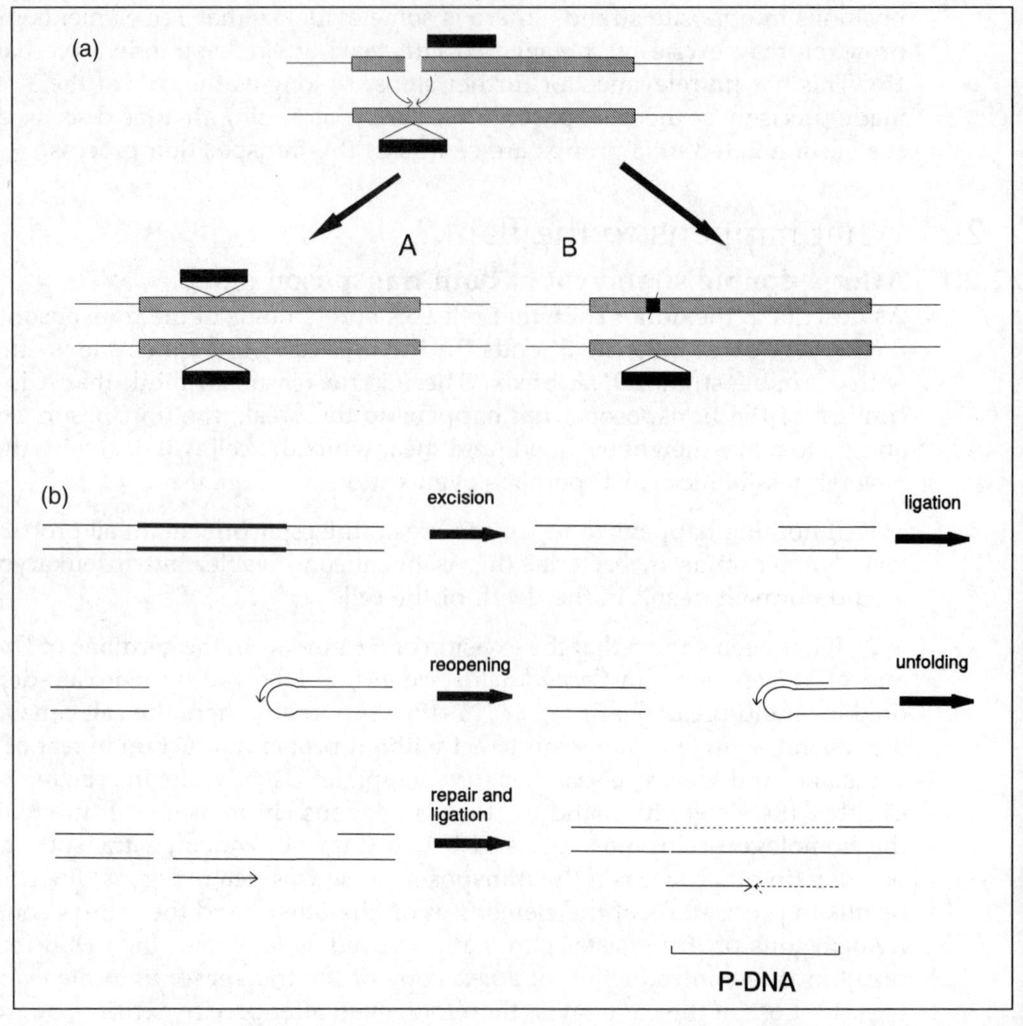

Fig. 3 (a) Transposon footprints as a result of imprecise DSB repair. Precise excision of a transposon followed by DSB repair which uses the sister chromatid or homologous chromosome as template usually results in re-instalment of the transposon sequences (option A). In rare cases the repair reaction is interrupted prematurely or is otherwise imprecise, which can result in deletions or only partial re-insertion of sequences encoded by the template DNA (option B). In genetic experimentation, option A will usually go unnoticed, since transposon excision followed by re-instalment will not result in change of DNA sequence. Therefore the relatively rare cases of imprecise repair will be what is noted. The subset of imprecise repair events that is noted depends on the restraints that phenotypic selection puts on the DNA sequence in the area of the transposon. If, for example, the transposon is in a coding region, then only footprints that leave sequences of lengths that are a multiple of three, and that do not contain stop codons, will be noticed as revertants of the phenotype caused by the initial transposon integration. For further discussion see refs 16 and 17. (b) Formation of P-DNA sequences. Through the process of hairpin formation by strand-to-strand ligation and subsequent re-opening through cutting at a different position, and ligation and repair, the sequence between the sites where the hairpin was closed and where it is opened becomes duplicated in opposite orientation. (See arrow to mark this sequence.) It is referred to as P-DNA (28). Note that whereas in Fig. 1 and in panel (a) a single line represents a double-stranded DNA molecule, it here represents a single DNA strand. The polarity of the strands is indicated: arrowheads are 3′ ends.

been found to be carried out by the same transposase, and since transposition insertion seems to be almost precise, it seems unnecessary to assume that excision is ever otherwise. The repair template does not have to be allelic, but can be ectopic (i.e. at another place in the genome), as long as it is sufficiently similar to the broken DNA (18, 19).

3. Alternatives for template-dependent DSB repair do exist. The two ends can be simply joined, as has been suggested to happen after somatic excision of Tc*1* (21), and excision of the Tec and IES elements in the rearrangement of the ciliate micronucleus into macronucleus (14, 15, 22). For bacterial Tn*10* it has been shown that simple rejoining of the broken donor DNA ends is probably not a common event (23).

4. Joining can be accompanied by deletion, as has been observed after excision of *Drosophila P* elements (24, 25) and Tc*1* elements of *C. elegans* (26, 27).

5. Several plant transposons exhibit unusual footprints, containing short inverted stretches that are similar to one or both ends of the flanking DNA. This has led Coen and co-workers to suggest that the broken ends initially form hairpin structures by direct ligation of the 5′ to the 3′ end of the same DNA end (28). Reopening of the hairpin at another position, followed by sealing and repair, would account for these types of footprints (see Fig. 3b). It should be emphasized that molecular support for this model has thus far not been obtained. However, support comes from an unexpected source: the coding joint in V–J recombination sometimes shows the same sequence configuration (P-DNA), and it has recently been shown that—at least in severe combined immunodeficiency (*scid*) mice, in which the formation of coding joints is seriously reduced—covalent links between the two strands at both coding ends occur (29).

2.2.2 After a single strand nick at the 3′ ends

After a nick at the 3′ ends of the prokaryotic transposons Mu, γδ, and Tn*3* and of retro-elements (such as HIV, Ty, Moloney murine leukaemia virus and Rous sarcoma virus) DNA sequences are left dangling at the 5′ ends, while the 3′-OH groups are available for strand transfer to target DNA. For retroviruses the remaining flank is only two nucleotides, and these can probably be removed by cellular repair enzymes, after the strand transfer reaction has been completed (see below). For Mu, Tn*3*, etc., the 5′ ends remain linked to the original flanking DNA, and therefore after the strand transfer (see below) the donor DNA has become physically linked to the target DNA. This implies that a subsequent reaction is now necessary to reconstitute the original configuration of the genome ('resolution', see below).

2.3 Strand transfer

Strand transfer is sometimes also referred to as 'joining', especially by retrovirologists, who associate 'strand transfer' with the template switching in reverse transcription mentioned above. In the strand transfer reaction the 3′-OH groups, released from flanking DNA by the donor cut reaction, are in their turn the nucleophiles that attack phosphodiester bonds in target DNA. Analysis of the stereochemical configuration of the phosphate group in the target DNA before and after the reaction is in agreement with a reaction where transposase mediates a one-step direct nucleophilic attack of these OH groups onto the target phosphates (12, 30). Note that in the strand transfer reaction two phosphodiester bonds are broken (in the target DNA), and two new bonds are made (between transposon and target DNA). Therefore, there is no net loss of high energy phosphodiester bonds: the direct transfer of target phosphate from flanking 3′-OH groups to transposon 3′-OH groups in one concerted reaction ensures that the reaction can proceed in the absence of energy carriers such as ATP. By analogy, it is reasonable to expect the same mechanism for Tn*10* integration (31; R. Chalmers and N. Kleckner, unpublished data), and *Drosophila P* element integration (32), both of which also take place in the absence of ATP.

Transposon integration involves not one but two trans-phosphorylation reactions: of each end of the transposon DNA to one strand of the target DNA and, since these are normally a few base-pairs apart, this results (after repair replication) in the target duplication characteristic for transposons. The integration-competent transposon is presumably in a DNA–protein complex, in which the two transposon ends are held together by transposase molecules, and possibly other factors. Such a configuration has been demonstrated for the Mu transposon (33, 34), Tn*10* (31), Tn*7* (35), and the 'core particle' of retro-elements (36; M. S. Lee and R. Craigie, unpublished data), and also has to be assumed for all other transposons to account for precise integration of both transposon ends into the same target DNA. The distance between the phosphodiester bonds in the two strands of the target DNA that are broken and joined to transposon ends is presumably determined by the precise structure of the 'transpososome'. Several related retroviruses exhibit different lengths of target site duplication. Since it has been shown that purified integrase can at low frequencies integrate both ends of a viral DNA analogue with the target duplication characteristic of its virus (37), it seems that the length of the staggered cut is an intrinsic property of the integrase protein. One could imagine that two monomers or oligomers of integrase each integrate one viral DNA end into different strands of target DNA, that the two ends are co-ordinated through protein–protein interactions between integrase subunits, and that the precise spacing between the two active sites is determined by this dimerization, and thus by the structure of integrase.

2.4 After the strand transfer

The products of the strand transfer reaction are summarized in Fig. 4b. In all cases, the newly integrated transposon is flanked by short stretches of single-stranded target DNA at each end. The cellular repair process that is presumably responsible for conversion of these stretches into double-stranded DNA creates the target site duplications that are characteristic for virtually all transposon insertion reactions. It is assumed, but not proven, that the few nucleotides present at retroviral 5′ ends (Fig. 4a) are removed during the repair process. Figure 4b shows the Mu strand transfer complex, which looks remarkably like a combination of two replication forks and is indeed subsequently replicated. When the donor is in one circular replicon, and the target in another, this results in the integration of one into the other, concomitant with transposition (hence the name 'co-integrate', see Chapter 5). To avoid these consequences, many transposons have employed a separate site-specific recombination system to resolve co-integrates. The model for co-integrate formation and resolution, now textbook material, was first suggested by Shapiro (38) and by Arthur and Sherratt (39), and the resolution reaction [catalysed by resolvase (or TnpR protein] was first demonstrated *in vitro* by Reed and Grindley (40). The mechanism of action of γδ and Tn*3* resolvase has been studied in much detail. Resolvase is one of the very few recombination enzymes whose three-dimensional structure is largely known (41). Resolution by site-specific recombination is reviewed in Chapter 5.

3. The transposase

Several lines of evidence indicate that donor cutting and strand transfer are catalysed by the same protein, the transposase. In some cases, direct evidence for this has been obtained; for example, retroviral integrase protein purified to homogeneity from recombinant DNA source exhibits both activities *in vitro* (37, 42) and so do Mu A protein (6) and Tn*10* transposase (13). Tn*7* transposases have been partially purified and found to be active *in vitro* (43). In *C. elegans* proficiency for transposon excision and for transposition correlate (strains that are active for one are active for the other) (44), and expression of transposase in a deficient strain activates both excision and transposition (45). This is in agreement with the assumption that both are carried out by the same enzymatic machinery [note that it has been possible to isolate mutants of Tn*10* transposase that are proficient for excision, and deficient for subsequent re-integration (46)]. It is not surprising that excision and integration are carried out by the same protein, and presumably even by the same active site on that protein, since the two reactions are mechanistically related. Donor cut, hydrolysis of a phosphodiester body, is a nucleophilic substitution attack of OH from water on a phosphate, and integration is the nucleophilic attack of OH of DNA sugar on a phosphate.

Mutational analysis of transposon ends indicates that several tens to several hundreds of base-pairs of each transposon end are required for transposition. It is reasonable to assume that transposases recognize these sequences and this has been observed for e.g. IS*903* (47), TnsB of Tn*7* (48), Mu A protein (33, 49), Tc*1* (50), and Tc*3* (45). Nevertheless, it has not yet been possible in all cases to detect specific binding of transposase to transposon ends. Several transposases (*P* transposase of *Drosophila*, and TnpA of maize *Ac*) bind to subterminal regions (51–53), which raises the question how precise cutting of transposon ends could be brought about

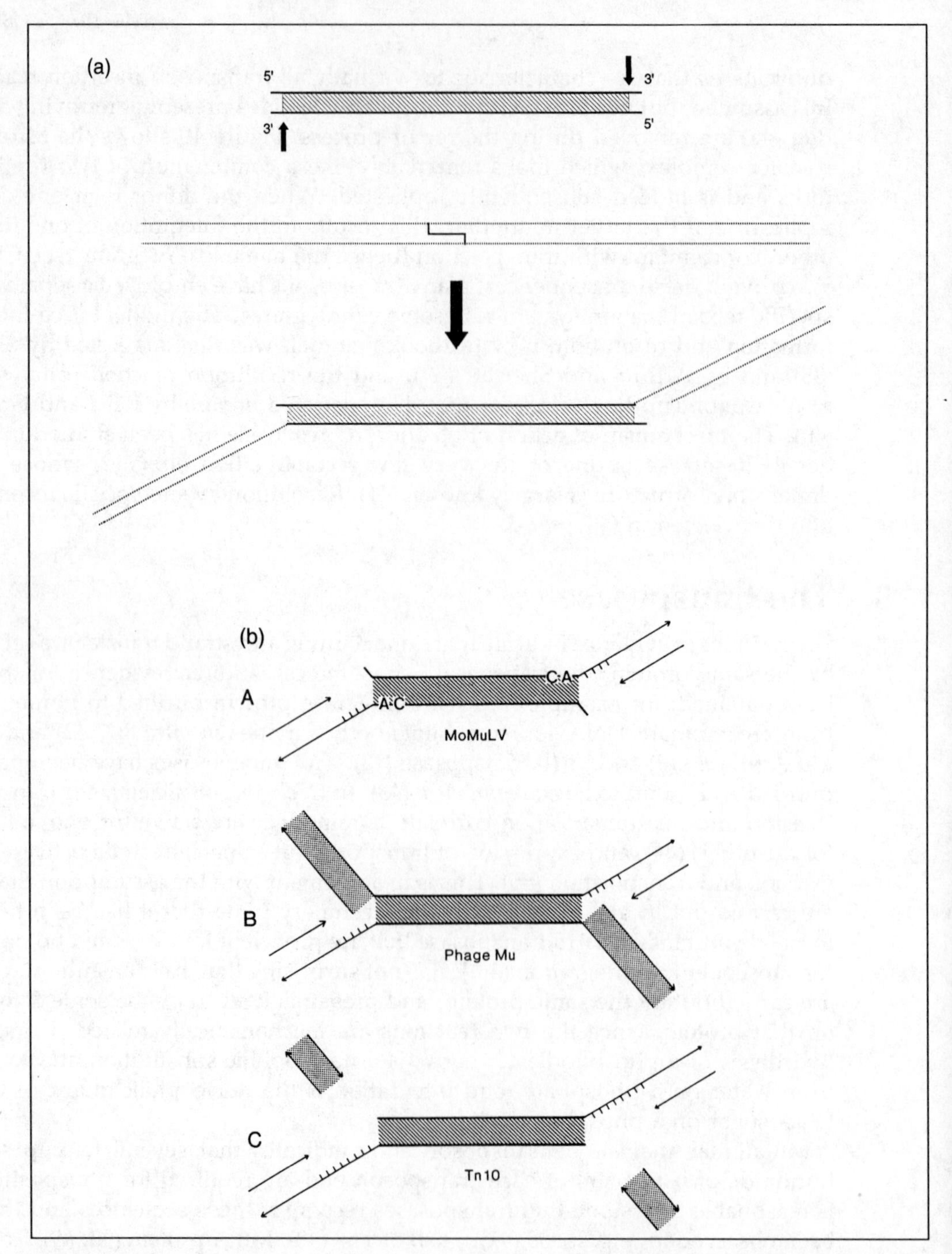

Fig. 4 Strand transfer complexes. (a) Donor cut followed by strand transfer reaction (here depicted for an integrating retrovirus). (b) Strand transfer complexes, the products of the strand transfer reactions. A = Moloney murine leukaemia virus, B = phage Mu, C = Tn*10*. Note that whereas in Figs 1 and 3a a single line represents a double-stranded DNA molecule, it here represents a single DNA strand. The polarity of the strands is indicated in panel (b): arrowheads are 3′ ends.

when the protein that mediates the cutting binds to a region that is a distance away from the transposon ends. Recent findings in our laboratory (50) are that the TcA transposase of nematode Tc*1* (Tc1A) contains an N-terminal region that specifically binds very near to transposon Tc*1* ends, but that the N-terminal region is apparently masked when it is present in the full-size transposase protein. The biological implications of these observations are not yet clear.

No complete transposase has been crystallized, and little is known about their three-dimensional structure. An obvious candidate for structural analysis is the HIV transposase IN, because of the perspective of structure-based drug design. The oligomeric composition of Mu transposase is a tetramer (53a).

Although all transposases do the same thing, they are not universally conserved in sequence. There is, however, some striking conservation of motifs between transposons of different phyla and regna. One class is formed by the transposases of *Drosophila hobo* elements, and plant *Ac*/*Ds* and Tam*3* elements (54). A second class is formed by the Tc family, found in different animals (nematodes, arthropods, and vertebrates) and protozoa (55–63). It has been suggested that the Tc class is more closely related to prokaryotic insertion elements such as IS*30* than to other eukaryotic transposons such as *P* elements (50). A striking observation is the similarity in the 'DD(35)E' motif of retroviral integrases and bacterial IS elements (64–66); the importance of the similarity gains credence from the observation that in a series of 43 mutations in virtually all conserved polar amino acids in HIV-2 IN, only the three 'DDE' positions were found to be essential for IN function (67), suggesting that the 'DDE' motif may be the core of the active site of both types of transposases.

In conclusion, the study of transposase proteins is just in its infancy, when compared, for example, with that of site-specific recombinases, and it will be interesting to see in detail how a single protein (with some help) can be responsible for the precise co-ordinated integration of two transposon ends into target DNA.

4. Host proteins and orientation selectivity

Most transposases do not function (well) without accessory ('host') factors. These accessory proteins seem to be responsible among other things for what is referred to as 'orientation selectivity' (sometimes called 'directionality'): the ability of the transposition machinery to sense whether the transposon ends are properly oriented, and only to act if they are. This has again been best documented for the Mu transposon: inversion of one transposon end results in loss of transpositional activity, and even of the first single strand nicks (donor cuts). In an elegant series of experiments Craigie and Mizuuchi (68) demonstrated that this long distance sensitivity is not brought about by a mechanism of one-dimensional sliding, gliding, or tracking of transposase between the two sites. They showed that the orientation could be sensed even when the two sites are on topologically intertwined separate circular DNA molecules, and thus cannot be the result of one-dimensional tracking. The sensing is apparently based on the structure of the synaptic complex, which is topologically restrained since it results from 'slithering' of intertwined double-stranded DNA molecules that are supercoiled. The synapsis is brought about around a site of 'condensation', the enhancer. This is where the accessory proteins come in: Mu A protein and host protein IHF bind the enhancer, and bend DNA at the positions where they bind (also see ref. 69). Note that free linear DNA molecules that contain an enhancer can also bring about synapsis (70); once synapsis is obtained, the enhancer can be experimentally removed and the reaction can nevertheless proceed. This demonstrates that the enhancer is only required for the establishment of synapsis (also see review by Boocock *et al.* (71)).

In addition to playing a role in bringing about synapsis, host proteins could play accessory roles in any of the other steps of transposition. In fact, since the transposition process involves a series of steps (synapsis, donor cut, strand transfer, break repair, replication and co-integrate resolution), it would not be surprising if accessory proteins played some role in many of these steps. In an elegant study yeast mutants were selected in which levels of Ty transposition were altered, and several complementation groups were defined. Some of these are real surprises, such as the *drb1* gene, apparently involved in RNA-debranching and Ty transposition (72).

5. Target choice

To the first approximation transposons insert in random sites in the target DNA. Nevertheless there are restrictions of several types and these are described here.

5.1 Target site or region preference

Alignment of target sites has in most cases not shown an obvious consensus sequence. An exception to this is the class of Tc*1*-like elements mentioned above (Section 3); these all integrate into the sequence TA (73). A possible additional consensus in the flanking region has also been proposed (74). Some of this possible additional non-randomness could be the result of non-randomness of TA sequences in the genome (they are, for example, more prevalent in introns than in exons).

The Tn*7* element has two distinct pathways, one for site-specific integration into an attachment site in the genome of Enterobacteriaceae, and one for random integration (75). The element encodes different transposases for the different reactions. It has recently been possible to reproduce the target-specific integration of Tn*7* *in vitro*, using four purified Tn*7*-encoded proteins (TnsA, TnsB, TnsC, and TnsD). The Tns A/B/C core is directed to its target, *attTn7*, by the TnsD protein (35).

A consensus preferred integration site has been derived for bacterial Tn*10* (76–78), and phage Mu integration is not completely random (79).

There is some evidence that in *Drosophila*, active genes are preferred targets for transposon insertion (80, 81). In yeast, promoter regions and tRNA genes provide 'hot' regions (82), and tRNA genes and LTR sequences are preferred for Ty*3* integration sites for Ty*1* (83).

Retrovirus integration exhibits some non-random patterns. A report that, of a few thousand independent retroviral isolates in mouse cells, a few were at precisely the same position of the genome (84) could not be confirmed by subsequent analysis (84a). *In vitro* studies suggest that nucleosome structure may to some extent be involved in determining non-randomness (85). This area has recently been reviewed (86).

A special case is the R2Bm element of the silk moth, *Bombyx mori*. This is one of a class of non-LTR retro-elements, found inserted at specific positions in the 28S rRNA genes of most insect species. Strictly speaking it is doubtful if this should be called a transposon, since its target site is always specific; on the other hand, the high number of rRNA genes, and thus of R2 target sites in the insect genomes, allows the element at least the 'choice' of which of these targets to occupy. R2Bm has been shown to encode its own integrase and reverse transcriptase (both functions carried out by the one encoded protein). In an elegant series of experiments, Luan *et al.* (87) recently showed that the target nick and reverse transcription step are coupled: the R2 protein nicks the target DNA in one strand, and uses the newly generated 3′-OH to prime reverse transcription on its genomic RNA. Then the other strand of the target is also cleaved; the result of this reaction is an integrated DNA–RNA hybrid. The second strand DNA synthesis could conceivably be carried out by host DNA polymerases. It remains to be determined whether this mechanism also applies to non-LTR retro-elements with more random target sites, such as the LINES and SINES that make up much of higher eukaryote genomes and fly I-factors (e.g. ref. 88; see discussion in ref. 86).

5.2 Regional re-integration

This phenomenon can only be studied in experimental systems where the donor and target sites of a single jump are known (which excludes situations where the genome is full of identical transposons). The phenomenon has been well documented for plant and *Drosophila* transposons (80, 81, 89–91): 10–20% of excised transposons integrate within a few kilobase pairs from the excision site. These transposons presumably jump via a cut-and-paste mechanism, i.e. through a 'free' intermediate. How can this freedom be reconciled with re-integration into the old neighbourhood? The preferred explanation seems to be that the donor and the target DNA meet before the transposon is excised, and that the excised element is therefore never really free (but only seems so in deproteinized purified DNA preparations). One could imagine an alternative explanation: diffusion of excised transposon plus proteins could be limited, for example because the transposon glides along DNA or chromatin, or because diffusion is otherwise partially limited to regions of DNA that are within one area of a chromosome (e.g. because the excised transposon is catenated with the chromosomal DNA).

5.3 *Cis*-immunity

Transposons need to integrate into host DNA, and not into themselves (since that would be suicidal). Those transposons that can spread horizontally between bacterial cells, by mobilization of the plasmids in which they are inserted, do better to transpose into another replicon than into a plasmid that already contains a transposon. The latter goal is achieved by what is called '*cis*-immunity', whereby the presence of one transposon in a plasmid makes the plasmid immune for integration of a second copy of the same element. The phenomenon is well-known for many transposons, and best analysed for phage Mu: transposon ends in the plasmid are entry points for Mu A transposase, which chases all Mu B protein that glides along the same DNA (Mu B marks the potential target site for integration for a Mu element) (92).

Retroviral genomic DNA does not integrate into itself, but readily integrates when it encounters foreign DNA. Experiments with partially purified, integration-competent complexes have shown that this self-immunity depends on reaction conditions (93). Recent experiments suggest that Moloney murine leukaemia virus prevents suicidal self-integration by the presence of factors that bind to the DNA between the ends, that can be removed or altered experimentally (by high salt treatment) without loss of integration capacity but with loss of self-immunity (M. S. Lee and R. Craigie, unpublished data).

6. Conclusion

Transposons are interesting 'forms of life' for several reasons. A discussion of their function quickly leads into the question what 'function' is in biology, and thus into evolutionary theory. They are interesting as tools, since they will do *in vivo* what molecular biologists have done for the last 25 years *in vitro*; namely combine DNA sequences. As illustrated above, they are also interesting from a mechanistic viewpoint, because they cover the intermediate position between complexity and simplicity, where meaningful biological problems can be understood in fairly simple mechanistic terms.

Acknowledgements

I am grateful to Sean Colloms, Henri van Luenen, Chris Vos, and Piet Borst for critical reading of the manuscript, and many useful suggestions, and to Nicole Immink for excellent preparation of the manuscript. I thank Bob Craigie, Ron Chalmers, and Nancy Kleckner for permission to cite unpublished work.

References

1. McClintock, B. (1984) The significance of responses of the genome to challenge. *Science*, **226**, 792.
2. Berg, D. E. and Howe, M. M. (eds) (1989) *Mobile DNA*. American Society for Microbiology, Washington, DC.
3. Plasterk, R. H. A. (1990) The ins and outs of transposition: molecular mechanisms of transposition and its control. *New Biol.*, **2**, 787.
4. Finnegan, D. J. (1992) Transposable elements. *Curr. Opin. Genet. Dev.*, **2**, 861.

5. Mizuuchi, K. (1992) Polynucleotidyl transfer reactions in transpositional DNA recombination. *J. Biol. Chem.*, **267**, 21273.
6. Mizuuchi, K. (1992) Transpositional recombination: mechanistic insights from studies of Mu and other elements. *Annu. Rev. Biochem.*, **61**, 1011.
7. Haniford, D. B. and Chaconas, G. (1992) Mechanistic aspects of DNA transposition. *Curr. Opin. Genet. Dev.*, **2**, 698.
8. Varmus, H. and Brown, P. (1989) Retroviruses. In *Mobile DNA*. D. E. Berg and M. M. Howe (eds). American Society for Microbiology, Washington, DC, p. 53.
9. Boeke, J. D. and Sandmeyer, S. B. (1991) Yeast transposable elements. In *The Molecular and Cellular Biology of the Yeast Saccharomyces*. J. Broach, E. Jones, and J. Pringles (eds). Cold Spring Harbor Laboratory Press, Cold Spring Harbor, NY, p. 193.
10. Eickbush, T. H. (1992) Transposing without ends: the non-LTR retro-transposable elements. *New Biol.*, **4**, 430.
11. Vink, C., Yeheskiely, E., van der Marel, G. A., van Boom, J. H., and Plasterk, R. H. A. (1991) Site-specific hydrolysis and alcoholysis of human immunodeficiency virus DNA termini by the viral integrase protein. *Nucleic Acids Res.*, **19**, 6691.
12. Engelman, A., Mizuuchi, K., and Craigie, R. (1991) HIV-1 DNA integration: mechanism of viral DNA cleavage and DNA strand transfer. *Cell*, **67**, 1211.
13. Benjamin, H. W. and Kleckner, N. (1992) Excision of Tn*10* from the donor site during transposition occurs by flush double-strand cleavage at the transposon termini. *Proc. Natl Acad. Sci. USA*, **89**, 4648.
14. Tausta, S. L. and Klobutcher, L. A. (1989) Detection of circular forms of eliminated DNA during macronuclear development in *E. crassus*. *Cell*, **59**, 1019.
15. Klobutcher, L. A. and Jahn, C. L. (1991) Developmentally controlled genomic rearrangements in ciliated protozoa. *Curr. Opin. Genet. Dev.*, **1**, 397.
16. Engels, W. R., Johnson-Schlitz, D. M., Eggleston, W. B., and Sved, J. (1990) High-frequency P element loss in *Drosophila* is homolog dependent. *Cell*, **62**, 515.
17. Plasterk, R. H. A. (1991) The origin of footprints of the Tc*1* transposon of *Caenorhabditis elegans*. *EMBO J.*, **10**, 1919.
18. Gloor, G. B., Nassie, N. A., Johnson-Schlitz, D. M., Preston C. R., and Engels, W. R. (1991) Targeted gene replacement in *Drosophila* via P element-induced gap repair. *Science*, **253**, 1110.
19. Plasterk, R. H. A. and Groenen, J. T. M. (1992) Targeted alterations of the *Caenorhabditis elegans* genome by transgene instructed DNA double strand break repair following Tc*1* excision. *EMBO J.*, **11**, 287.
20. Haring, M. A., Scofield, S., Teeuwen, M. J., Leuving, G. S., Nijkamp, H. J. J., and Hille, J. (1991) Novel DNA structures resulting from dTam3 excision in tobacco. *Plant Mol. Biol.*, **17**, 995.
21. Emmons, S. W. and Yesner, L. (1984) High-frequency excision of transposable element Tc*1* in the nematode *Caenorhabditis elegans* is limited to somatic cells. *Cell*, **36**, 599.
22. Tausta, S. L., Turner, L. A., Buckley, L. K., and Klobutcher, L. A. (1991) High fidelity developmental excision of Tec1 transposons and internal eliminated sequences in *Euplotes crassus*. *Nucleic Acids Res.*, **19**, 3229.
23. Bender, J., Kuo, J., and Kleckner, N. (1991) Genetic evidence against intramolecular rejoining of the donor DNA molecule following IS*10* transposition. *Genetics*, **128**, 687.
24. Tsubota, S. and Schedl, P. (1986) Hybrid dysgenesis-induced revertants of insertions at the 5′ end of the rudimentary gene in *Drosophila melanogaster*: transposon-induced control mutations. *Genetics*, **114**, 165.

25. Salz, H. K., Cline, T. W., and Schedl, P. (1987) Functional changes associated with structural alterations induced by mobilization of a *P* element inserted in the sex-lethal gene of *Drosophila*. *Genetics*, **117,** 221.
26. Kiff, J. E., Moerman, D. G., Schriefer, L. A., and Waterston, R. H. (1988) Transposon-induced deletions in *unc-22* of *C. elegans* associated with almost normal gene activity. *Nature*, **331,** 631.
27. Zwaal, R. R., Broeks, A., Van Meurs, J., Groenen, J. T. M., and Plasterk, R. H. A. (1993) Target-selected gene inactivation in *Caenorhabditis elegans*, using a frozen transposon insertion mutant bank. *Proc. Natl Acad. Sci. USA*, in press.
28. Coen, E. S., Robbins, T. P., Almeida, J., Hudson, A., and Carpenter, R. (1989) Consequences and mechanism of transposition in *Antirrhinum majus*. In *Mobile DNA*. D. E. Berg and M. M. Howe (eds). American Society for Microbiology, Washington, DC, p. 413.
29. Roth, D. B., Menetski, J. P., Nakajima, P. B., Bosma, M. J., and Gellert, M. (1992) V(D)J recombination: broken DNA molecules with covalently sealed (hairpin) coding ends in scid mouse thymocytes. *Cell*, **70,** 983.
30. Mizuuchi, K. and Adzuma, K. (1991) Inversion of the phosphate chiralty at the target site of Mu DNA strand transfer: evidence for a one-step transesterification mechanism. *Cell*, **66,** 129.
31. Haniford, D. B., Benjamin, H. W., and Kleckner, N. (1991) Kinetic and structural analysis of a cleaved donor intermediate and a strand transfer intermediate in Tn*10* transposition. *Cell*, **64,** 171.
32. Kaufman, P. D. and Rio, D. C. (1992). *P* element transposition *in vitro* proceeds by a cut-and-paste mechanism and uses GTP as a cofactor. *Cell*, **69,** 27.
33. Craigie, R., Mizuuchi, M., and Mizuuchi, K. (1984) Site-specific recognition of the bacteriophage Mu ends by the Mu A protein. *Cell*, **39,** 387.
34. Nakayama, C., Teplow, D. P., and Harshey, R. M. (1987) Structural domains in phage Mu transposase: identification of the site-specific DNA binding domain. *Proc. Natl Acad. Sci. USA*, **84,** 1809.
35. Bainton, R. J., Kubo, K. M., Feng, J., and Craig, N. L. (1993) Tn7 transposition: target DNA recognition is mediated by multiple Tn7-encoded proteins in a purified *in vitro* system *Cell*, **72,** 931.
36. Farnet, C. M. and Haseltine, W. A. (1991) Determination of viral proteins present in the human immunodeficiency virus type 1 preintegration complex. *J. Virol.*, **65,** 1910.
37. Craigie, R., Fujiwara, T., and Bushman, F. (1990) The IN protein of Moloney murine leukemia virus processes the viral DNA ends and accomplishes their integration *in vitro*. *Cell*, **62,** 829.
38. Shapiro, J. A. (1979) Molecular model for the transposition of bacteriophage Mu and other transposable elements. *Proc. Natl Acad. Sci. USA*, **76,** 1933.
39. Arthur, A. and Sherratt, D. (1979) Dissection of the transposition process: a transposon-encoded site-specific recombination system. *Mol. Gen. Genet.*, **175,** 267.
40. Reed, R. R. and Grindley, N. D. F. (1981) Transposon-mediated site-specific recombination *in vitro*: DNA cleavage and protein–DNA linkage at the recombination site. *Cell*, **25,** 721.
41. Sanderson, M. R., Freemont, P. S., Rice, P. A., Goldman, A., Hatfull, G. F., Grindley, N. D. F. and Steitz, T. A. (1990) The crystal structure of the catalytic domain of the site-specific recombination enzyme γδ resolvase at 2.7 Å resolution. *Cell*, **63,** 1323.
42. Katz, R. A., Merkel, G., Kulkolsky, J., Leis, J., and Skalka, A. M. (1990) The avian

retroviral IN protein is both necessary and sufficient for integrative recombination *in vitro*. *Cell*, **63,** 87–95.

43. Bainton, R. J., Gamas, P., and Craig, N. L. (1991). Tn*7* transposition *in vitro* proceeds through an excised transposon intermediate generated by staggered breaks in DNA. *Cell*, **65,** 805.
44. Collins, J., Saari, B., and Anderson, P. (1987) Activation of a transposable element in the germ line but not the soma of *Caenorhabditis elegans*. *Nature*, **328,** 726.
45. Van Luenen, H. G. A. M., Colloms, S. D., and Plasterk, R. H. A. (1993) Mobilization of quiet, endogenous Tc3 transposons of *Caenorhabditis elegans* by forced expression of Tc3 transposase. *EMBO J.*, **12,** 2513.
46. Haniford, D. B., Chelouche, A. R., and Kleckner, N. (1989) A specific class of IS*10* transposase mutants are blocked for target site interactions and promote formation of an excised transposon fragment. *Cell*, **59,** 385.
47. Derbyshire, K. M. and Grindley, N. D. F. (1992) Binding of the IS*903* transposase to its inverted repeat *in vitro*. *EMBO J.*, **11,** 3449.
48. Tang, Y., Lichtenstein, C., and Cotterill, S. (1991) Purification and characterisation of the TnsB protein of Tn*7*: a transposition protein that binds to the ends of Tn*7*. *Nucleic Acids Res.*, **19,** 3395.
49. Leung, P. C., Teplow, D. B., and Harshey, R. M. (1989) Interaction of distinct domains in Mu transposase with Mu DNA ends and an internal transpositional enhancer. *Nature*, **338,** 656.
50. Vos, J. C., van Luenen, H. G. A. M., and Plasterk, R. H. A. (1993) Characterization of the *Caenorhabditis elegans* Tc*1* transposase *in vivo* and *in vitro*. *Genes Dev.*, **7,** 1244.
51. Rio, D. R. and Rubin, G. M. (1988) Identification and purification of a *Drosophila* protein that binds to the terminal 31-base-pair inverted repeats of the *P* transposable element. *Proc. Natl Acad. Sci. USA*, **85,** 8929.
52. Kaufman, P. D., Doll, R. F., and Rio, D. C. (1989) *Drosophila P* element transposase recognizes internal *P* element DNA sequences. *Cell*, **59,** 359.
53. Kunze, R. and Starlinger, P. (1989) The putative transposase of transposable element *Ac* from *Zea mays* L. interacts with subterminal sequences of *Ac*. *EMBO J.*, **8,** 3177.

53a. Baker, T. A. and Mizuuchi, K. (1992) DNA-prompted assembly of the active tetramer of the Mu transposase. *Genes Dev.*, **6,** 2221.

54. Calvi, B. R., Hong, T. J., Findley, S. D., and Gelbart, W. M. (1991) Evidence for a common evolutionary origin of inverted repeat transposons in *Drosophila* and plants: *hobo, Activator* and Tam3. *Cell*, **66,** 465.
55. Henikoff, S. and Plasterk, R. H. A. (1988) Related transposons in *C. elegans* and *D. melanogaster*. *Nucleic Acids Res.*, **16,** 6234.
56. Harris, L. J., Baillie, D. L., and Rose, A. M. (1988) Sequence identity between an inverted repeat family of transposable elements in *Drosophila* and *Caenorhabditis* *Nucleic Acids Res.*, **16,** 5991.
57. Prasad, S. S., Harris, L. J., Baillie, D. L., and Rose, A. M. (1991) Evolutionary conserved regions in *Caenorhabditis* transposable elements deduced by sequence comparison. *Genome*, **34,** 6.
58. Brezinsky, L., Wang, G. V. L., Humphreys, T., and Hunt, J. (1990) The transposable element Uhu from Hawaiian *Drosophila*—member of the widely dispersed class of Tc*1*-like elements. *Nucleic Acids Res.*, **18,** 2053.
59. Franz, G. and Savakis, C. (1992) Minos, a new transposable element from *Drosophila hydei*, is a member of the Tc*1*-like family of transposons. *Nucleic Acids Res.*, **19,** 6646.

60. Abad, P., Quiles, C., Tares, S., Piotte, C., Castagnone-Sereno, P., Abadon, M., and Dalmasso, A. (1991) Sequences homologous to Tc(s) transposable elements of *Caenorhabditis elegans* are widely distributed in the phylum Nematoda. *J. Mol. Evol.*, **33,** 251.
61. Heierhorst, J., Lederis, K., and Richter, D. (1992) Presence of a member of the Tc*1*-like transposon family from nematodes and *Drosophila* within the vasotocin gene of a primitive vertebrate, the Pacific hagfish *Eptatretus stouti*. *Proc. Natl Acad. Sci. USA*, **89,** 6798.
62. Henikoff, S. (1992) Detection of *Caenorhabditis* transposon homologs in diverse organisms. *New Biol.*, **4,** 382.
63. Caizzi, R., Caggese, C., and Pimpinella, S. (1993) Bari-1, a new transposon-like family in *Drosophila melanogaster* with a unique heterochromatic organization. *Genetics*, **133,** 335.
64. Fayet, O., Ramond, P., Polard, P., Prère, M. F., and Chandler, M. (1990) Functional similarities between retroviruses and the IS3 family of bacterial insertion sequences. *Mol. Microbiol.*, **4,** 1771.
65. Khan, E., Mack, J. P. G., Katz, R. A., Kulkosky, J., and Skalka, A. M. (1991) Retroviral integrase domains: DNA binding and the recognition of LTR sequences. *Nucleic Acids Res.*, **19,** 851.
66. Kulkosky, J., Jones, K. S., Katz, R. A., Mack, J. P. G., and Skalka, A. M. (1992) Residues critical for retroviral integrative recombination in a region that is highly conserved among retroviral/retrotransposon integrases and bacterial insertion sequence transposases. *Mol. Cell. Biol.*, **12,** 2331.
67. Van Gent, D., Oude Groeneger, A. A. M., and Plasterk, R. H. A. (1992) Mutational analysis of HIV-2 integrase. *Proc. Natl Acad. Sci. USA*, **89,** 9598.
68. Craigie, R. and Mizuuchi, K. (1986) Role of DNA topology in Mu transposition: mechanism of sensing the relative orientation of two DNA segments. *Cell*, **45,** 793.
69. Johnson, R. C. and Simon, M. I. (1987) Enhancers of site-specific recombination in bacteria. *Trends Genet.*, **3,** 262.
70. Surette, M. G. and Chaconas, G. (1992) The Mu transpositional enhancer can function *in trans*: requirement of the enhancer for synapsis but not strand cleavage. *Cell*, **68,** 1101.
71. Boocock, M. R., Rowland, S. J., Stark, W. M., and Sherrat, D. J. (1992) Insistent and intransigent: a phage Mu enhancer functions *in trans*. *Trends Genet.*, **8,** 151.
72. Chapman, K. B. and Boeke, J. D. (1991) Isolation and characterization of the gene encoding yeast debranching enzyme. *Cell*, **65,** 483.
73. Rosenzweig, B., Liao, L. W., and Hirsch, D. (1963) Target sequences for the *C. elegans* transposable element Tc1. *Nucleic Acids Res.*, **11,** 7137.
74. Mori, I., Benian, G. M., Moerman, D. G., and Waterston, R. H. (1988) Transposable element Tc*1* of *Caenorhabditis elegans* recognizes specific target sequences for integration. *Proc. Natl Acad. Sci. USA*, **85,** 861.
75. Craig, N. (1989) Transposon Tn7. In *Mobile DNA*. D. E. Berg and M. M. Howe (eds). American Society for Microbiology, Washington, DC, p. 211.
76. Kleckner, N., Steele, D., Reichardt, K., and Botstein, D. (1979) Specificity of insertion by the translocatable tetracycline-resistance element Tn*10*. *Genetics*, **92,** 1023.
77. Kleckner, N. (1989) Transposon Tn*10*. In *Mobile DNA*. D. E. Berg and M. M. Howe (eds). American Society for Microbiology, Washington, DC, p. 229.
78. Bender, J. and Kleckner, N. (1992) Tn*10* insertion specificity is strongly dependent upon sequences immediately adjacent to the target-site consensus sequence. *Proc. Natl Acad. Sci. USA*, **89,** 7996.

79. Raiband, O., Roa, M., Braun-Breton, and Schwartz, M. (1979) Structure of the *mal B* region in *E. coli* K12. *Mol. Gen. Genet.*, **174,** 241.
80. Tsubota, S., Ashburner, M., and Schedl, P. (1985) *P* element-induced control mutations at the *r* gene of *Drosophila melanogaster*. *Mol. Cell. Biol.*, **5,** 2567.
81. Kelly, M. R., Kid, S., Berg, R. L., and Young, M. W. (1987) Restriction of *P* element insertions at the *Notch* locus of *Drosophila melanogaster*. *Mol. Cell. Biol.*, **7,** 1545.
82. Sandmeyer, S. B., Bilanchone, V. W., Clark, D. J., Morcos, P., Carle, G. F., and Brodeur, G. M. (1988) Sigma elements are position specific for many different yeast tRNA genes. *Nucleic Acids Res.*, **16,** 1499.
83. Ji, H., Moore, D. P., Blomberg, M. A., Braiterman, L. T., Voytao, D. F., Natsoulis, G., and Boeke, J. D. (1993) Hotspots for unselected Ty1 transposition events on yeast chromosome III are near tRNA genes and LTR sequences. *Cell*, **73,** 1007.
84. Shih, C. C., Stoye, J. P., and Coffin, J. M. (1988) Highly preferred targets for retrovirus integration. *Cell*, **53,** 531.
84a. Withers-Ward, E. S., Kitamura, Y., Barnes, J. P., Coffin, J. M., (1994) Distribution of targets for avian retrovirus DNA integration *in vivo*. *Genes Dev.*, **8,** 1473.
85. Pryciak, P. M. and Varmus, H. E. (1992) Nucleosomes, DNA binding proteins, and DNA sequence modulate retroviral integration target site selection. *Cell*, **69,** 769.
86. Craigie, R. (1992) Hot spots and warm spots: integration specificity of retroelements. *Trends Genet.*, **8,** 187.
87. Luan, D. D., Korman, M. H., Jakubczak, J. L., and Eickbush, T. M. (1993) Reverse transcription of R2Bm RNA is primed by a nick at the chromosomal target site: a mechanism for non-LTR retrotransposition. *Cell*, **72,** 595.
88. Fawcett, D. H., Lister, C. K., Kellett, E., and Finnegan, D. (1986) Transposable elements controlling I–R hybrid dysgenesis in *D. melanogaster* are similar to mammalian LINEs. *Cell*, **47,** 1007.
89. Moreno, M. A., Chen, J., Greenblatt, I., and Dellaporta, S. L. (1992) Reconstitutional mutagenesis of the maize *P* gene by short-range *Ac* transpositions. *Genetics*, **131,** 939.
90. Tower, J., Karpen, G. M., Craig, N., and Spradling, A. C. (1993) Preferential transposition of *Drosophila P* elements to nearby chromosomal sites. *Genetics*, **133,** 347.
91. Zhang, P. and Spradling, A. C. (1993) Efficient and dispersed local *P* element transposition from *Drosophila* females. *Genetics*, **133,** 361.
92. Adzuma, K. and Mizuuchi, K. (1988) Target immunity of Mu transposition reflects a differential distribution of Mu B protein. *Cell*, **53,** 257.
93. Ha Lee, Y. M. and Coffin, J. M. (1990) Efficient autointegration of avian retrovirus DNA *in vitro*. *J. Virol.*, **64,** 5958.

3 | Transposable elements as tools for molecular analyses in bacteria

CLAIRE M. BERG and DOUGLAS E. BERG

1. Introduction

The transposable elements of bacteria vary in size and structure, and in the ways they move. Some half-dozen elements and their engineered derivatives have been exploited as research tools since the late 1970s. Their use in conjunction with recombinant DNA methods has revolutionized the practice of microbial genetics. The value of transposons as molecular genetic tools stems from their ability to move as discrete DNA segments to new locations where they disrupt the target DNA. Natural and engineered transposons can serve as mobile sources of a variety of phenotypic or physical markers, including selectable (or sometimes counter-selectable) genes, predetermined restriction sites, reporter genes, promoters, and replication origins.

The requirements for transposition are very simple—just a short specific sequence (<20 bp in some cases) at each end of the transposon, an element-specific transposase, and host-encoded proteins. Virtually any DNA segment can be rendered mobile by placing it between transposon ends. This has allowed the construction of elements that are optimized for specific tasks such as mutagenesis, monitoring transcription or translation and protein localization (reporter transposons), induction of RNA synthesis (promoter transposons), and, more recently, DNA sequencing.

Much of the early impetus for use of bacteriophages λ and Mu, and antibiotic-resistance transposons such as Tn*5* and Tn*10*, came from the ease with which they generated single, loss-of-function chromosomal mutations, and the ease of detecting and characterizing previously unknown or poorly understood target genes marked by their insertion. The construction of a phage Mu derivative that contained a promoterless *lac* reporter gene by Casadaban and Cohen (1) opened up avenues of functional gene analysis that continue to be exploited to great advantage. The resistance genes present in most transposons allow target genes to be identified, mapped, moved into new strain backgrounds, and cloned much more efficiently than had been possible previously. Unique transposon sequences and restriction sites constitute molecular markers for physical mapping (2, 3). In the last decade, transposons have been used extensively to analyse DNA cloned in *Escherichia coli*, and also for molecular genetic analysis and manipulation in other Gram-negative and Gram-positive bacteria, as well as in eukaryotes. Tn*5*, in particular, has been widely used because it has a broad host range and transposes quite randomly in many Gram-negative bacterial species, although Tn*916* has been more useful than Tn*5* for *Neisseria* and *Haemophilus*. Tn*916* and the unrelated transposon Tn*917* have been most widely useful for Gram-positive bacteria.

Most insertions occur at sites that do not change the phenotype. However, they may be linked to genes of interest, particularly ones that cannot be selected directly. The insertions serve as easily selected genetic markers that facilitate gene mapping, strain construction, and localized mutagenesis. This is exemplified by a set of *E. coli* strains with transposon insertions marked with genes for tetracycline (Tet^r) or kanamycin (Kan^r) resistance approximately every 1 min (~50 kb)(4).

Current work with transposable elements follows three general tracks:

- the development of transposons and delivery systems for mutagenizing the chromosomes of an increasing array of bacterial species
- the development of transposons for isolating insertions in genes cloned in plasmid or phage vectors in *E. coli*
- the development of transposon-based cloning vectors for isolating nested deletions

In some cases these elements are constructed for doing reverse genetics, in which they insert into the cloned gene in *E. coli*, and are then introduced into the original host, where recombination into the chromosome can be selected and phenotypic studies undertaken.

A number of broad host range plasmids have been engineered to serve as shuttle vectors that move cloned DNA between different bacterial genera (5). The use of transposons expanded quickly during the 1980s, especially when recombinant DNA methods made it easier to construct special-purpose elements for a variety of tasks, including

- gene localization
- identifying open reading frames
- characterizing operon organization
- characterizing transcriptional and translational regulation
- determining cellular versus extracellular protein localization and probing protein conformation
- moving specific genes or sites into new hosts or to specific genomic locations
- cloning bacterial DNA *in vivo* (rather than *in vitro*)
- generating rearrangements, including deletions and replicon fusions
- DNA sequencing

This review will focus on how the mechanism of transposition dictates the choice of transposon system, and on the uses of engineered derivatives of Tn*5*, γδ, and Mu to facilitate the molecular analysis of DNAs cloned in *E. coli*. Some other useful transposons and their engineered derivatives will also be described. More information about transposons and their uses can be found in refs 3 and 6–8. The following reviews discuss important aspects of particular valuable transposons: γδ (9–11), Tn*5*/IS*50* (12–14), Tn*10*/IS*10* (15, 16), Tn*916* (17), Tn*917* (18), and phage Mu (19, 20).

1.1 Overview of transposition mechanism(s) and products

Certain basic features of transposition appear to be shared by all transposons, although the detailed mechanisms and the types, frequencies, and distributions of transposition products, are element-specific. In all transposons, the element-specific transposase protein recognizes a pair of short DNA sequences (often inverted repeats of 18–35 bp) at the transposon termini, cleaves one or both strands at both vector–transposon junctions, cleaves both strands of the target DNA, and joins the transposon and target together. In transposition from one molecule to another (intermolecular transposition), cleavage of both strands at each vector–transposon junction yields a simple insertion (conservative transposition) (Fig. 1, left), while cleavage of only one strand at each junction yields a co-integrate molecule (replicative transposition) (Fig. 1, right). Most elements, like one class of restriction enzymes, seem to cleave the target at staggered sites, leaving short 5′ extensions (usually 5 or 9 bp). Joining of the 3′ ends of the transposon to the 5′ ends of target strands leaves single-stranded gaps, which are probably filled by DNA repair (21; see also Chapter 2). In conservative, 'cut-and-paste' transposition (Tn*5* and Tn*10*), such repair synthesis yields the characteristic direct duplications that bracket most transposon insertions. The donor molecule is presumably destroyed, and its place is taken by sibling molecules (12) (Fig. 1, left).

In replicative transposition, the product is a co-integrate in which the donor and target molecules are joined by direct repeats of the element. Formation of this co-integrate entails gap-filling as above, and then continued DNA synthesis through the transposon, duplicating it and fusing the two parental replicons. A second step, which is usually considered part of the transposition process (but is not necessary, and is absent in intramolecular transposition), is the breakdown (resolution) of the co-integrate into an unchanged donor molecule and a target molecule that contains one copy of the transposon (Fig. 1, right), and is also bracketed by a short target site duplication. Resolution of the co-integrate is usually accomplished by a transposon-encoded, site-specific 'resolvase' acting on the transposon resolution site; it can also be accomplished by host generalized recombination.

The final product of both replicative and conservative intermolecular transposition is normally a simple insertion of the transposon into the target DNA. The elegant demonstrations of the replicative transposition model for Tn*3* (22) led to the simplifying assumption in some quarters that all transposition is replicative.

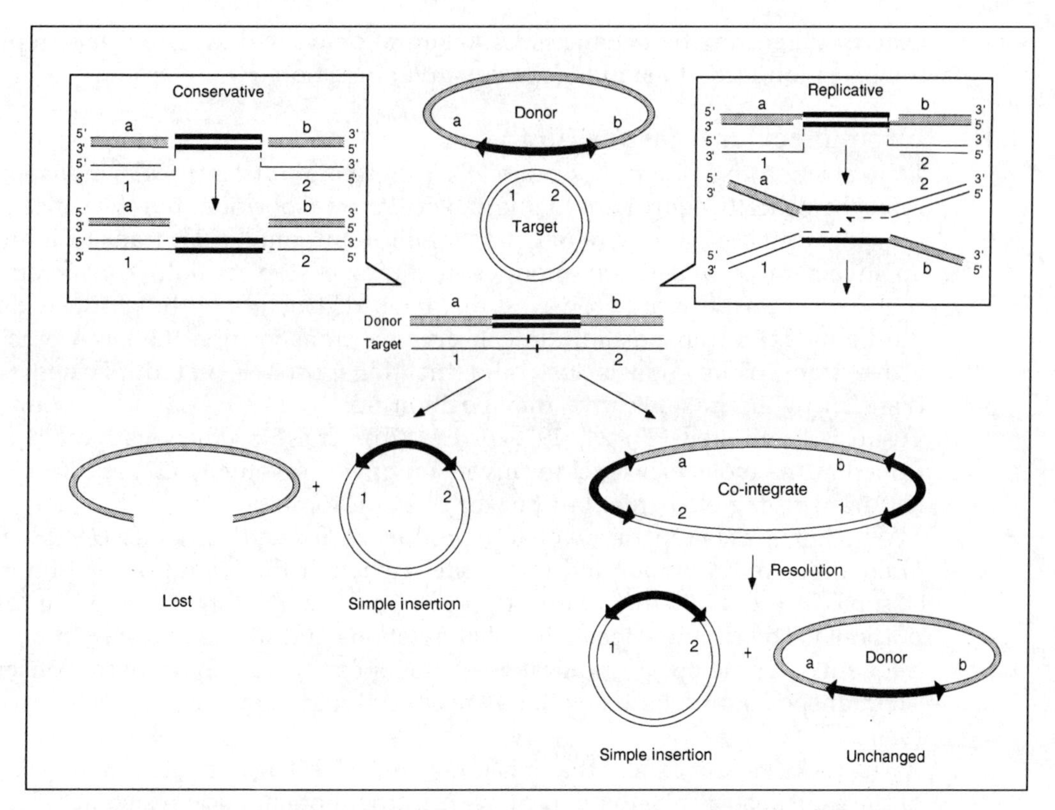

Fig. 1 Model for the formation of simple insertions by conservative and replicative intermolecular transposition. Left: conservative transposition, in which the donor molecule is believed to be destroyed by the transposition process (apparently unchanged donor molecules found after transposition are probably due to hyper-replication of sibling molecules) (12). Right: replicative transposition, in which the co-integrate product is resolved in a second recombination step (either site-specific or generalized) (11, 125).

However, true co-integrate replicative transposition products of conservatively transposing elements, such as Tn*5*, have not been described (see ref. 12 for a discussion of apparently conflicting results). Transposons that appear to yield only conservative transposition products include Tn*5*, Tn*7*, Tn*10*, Tn*916*, and also the initial integration of phage Mu after infection. In contrast, replicatively transposing elements, such as γδ and other members of the Tn*3* family, Tn*9*/*IS1*, and phage Mu (during lytic growth), yield a small fraction of conservative transposition products (12, 16, 17, 21, 23–26).

Are conservative and replicative transposition alternative aspects of the same basic process, or are they fundamentally different? *In vitro* studies suggest that transposase–transposon binding and junction cleavage are similar, and therefore that these pathways may be related (summarized in ref. 21; see also Chapter 2). In addition, the rare conservative transposition of γδ exhibits the same target site specificity as normal replicative transposition (G. Wang and C. M. Berg, unpublished), suggesting that chance second strand cleavage during normally replicative transposition leads to simple (conservative) insertion.

1.1.1 Intramolecular transposition

Transposition to a site in the same DNA molecule (intramolecular transposition), is mechanistically equivalent to intermolecular transposition, but the consequences are quite different, and generally less familiar. Intramolecular transposition yields an inversion or a deletion product in a single step in both conservative and replicative transposition. However, the detailed structure of the products depends on the mode of transposition: in conservative transposition the DNA outside the active transposon ends is lost, and the transposon is not duplicated (27, 28) (Fig. 2a), while in replicative transposition no DNA is lost and the entire transposon is duplicated (23, 24, 29, 30) (Fig. 2b). These predictions have been confirmed at the molecular level for inversion of conservatively (27) and replicatively (24) transposing elements, but not for deletion formation.

Whether a deletion or inversion product is formed depends on the relative orientations of the donor and target sites (whether the 3′ end of the donor transposon is inserted into the same or opposite strand of the target DNA) (Fig. 2, boxed portions). The simple expectation that deletions and inversions should be equally frequent, is borne out for replicative γδ transposition (24), but not for conservative Tn*5* transposition, where deletions are several-fold more frequent than inversions (27).

The lucky circumstances that it is possible to select against the functional *tet* gene of the transposon Tn*10* (by penicillin enrichment or chemical resistance: 31, 32) led to the widespread exploitation of intramolecular transposition to isolate deletions adjacent to Tn*10* chromosomal insertion sites, removing the *tet* gene while retaining an IS*10* element.

It should be noted that intramolecular conservative transposition involves the inner, rather than the outer IS element ends of these composite elements, and that the central portion of the transposon is lost. These active ends are inserted into a new site. This is shown clearly in inversion derivatives in which both products are recovered (the reciprocal deletion derivatives cannot be recovered because one lacks a replication origin and is lost). Inversion derivatives are smaller than the parental replicon because of loss of the central portion (27). The expectation that an element containing only a pair of inside transposon ends undergoes intramolecular transposition to generate inversions and deletions of the expected structures has been confirmed (27).

As described below, deletion products formed by intramolecular replicative transposition are very useful for generating restriction maps and ordered templates for sequencing because both DNA strands can be accessed in a single plasmid (9, 33).

1.1.2 Inverse transposition

Inverse transposition (34, 35) is the intermolecular analogue of conservative intramolecular transposition. The two inside ends of the IS elements in the composite

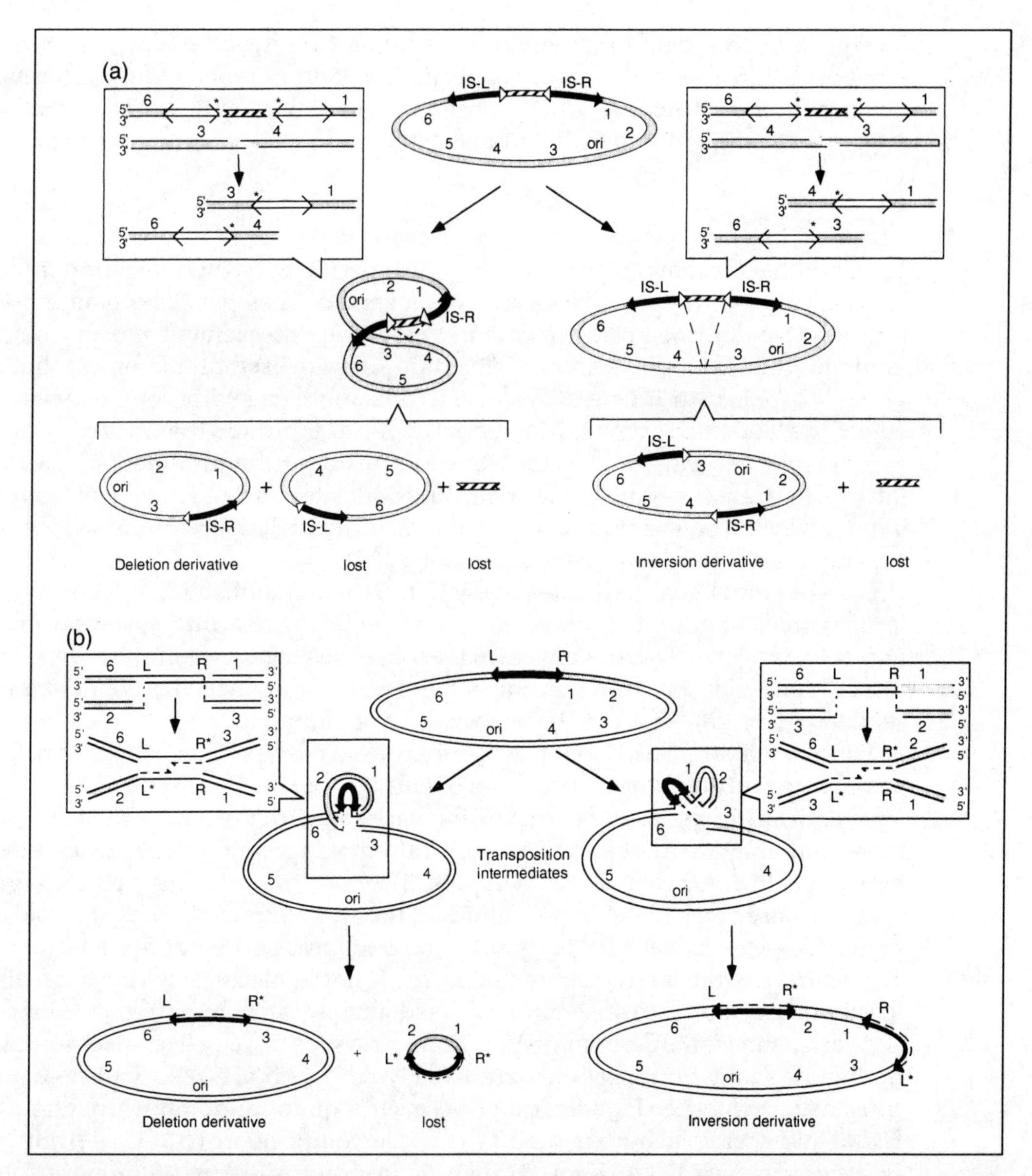

Fig. 2 Model for the formation of deletion and inversion products by conservative and replicative intramolecular transposition. (a) Conservative transposition, in which the 'donor molecule' is the segment marked by diagonal lines. This segment is not found in transposition products (27; C. Chin and C. M. Berg, in preparation). (b) Replicative transposition, in which the transposon is duplicated and no DNA is lost during transposition (one deletion product is generally lost because it lacks a replication origin) (24, 29).

transposon participate to move the entire donor replicon, except the central portion of the transposon, to a new replicon. This kind of replicon fusion forms a 'cointegrate' joint structure whose structure, however, differs from that of cointegrates formed by replicative transposition.

1.2 Target specificity

Transposable elements differ greatly in the specificity of their insertion into target DNAs. Tn*5* and γδ typically insert into dozens of sites per gene, and are among the most random of known elements. However, no element moves completely randomly. For example, some Tn*5* insertion sites are used preferentially (hotspots). These hotspots bear little resemblance to one another, except for GC pairs at each end of the 9 bp duplications. Mutational analyses indicated that recurrent insertion at a hotspot depends on several features, including the terminal GC pairs at the target site, transcriptional activity through the site, and negative DNA supercoiling. However, the factors that direct the majority of Tn*5* insertions, which are not into hotspots, are not yet well understood (12).

Tn*10* has also been used extensively to isolate mutations in *E. coli* and *Salmonella typhimurium*, although it tends to insert preferentially into just one or a few hotspots per gene. There is a weak 6 bp consensus sequence in the 9 bp target site duplication, with the choice of hotspot also being influenced by the flanking DNA sequence (16, 28). A Tn*10* transposase gene mutation that reduces insertional specificity making Tn*10* about as random as wild-type Tn*5*, offers considerable promise for further work with this element (15, 36).

γδ and Tn*3* transpose poorly into the bacterial chromosome, but seem to move more randomly than other transposons into most plasmid DNAs, usually without hotspots. However, occasional hotspots (37), and 'cold' regions that contain no or very few insertions (38–40), are found. Although not understood, the cold region pattern appears to be a property of an interaction between donor and target, not a property of the target alone, because donor molecules with γδ at different locations are found to insert into a 'cold region' at higher frequencies (X. Xu, G. Wang, and C. M. Berg, in preparation). The sites of γδ or Tn*3* insertion, defined by the 5 bp target duplications, are usually A+T-rich (41, 42), with no detectable consensus sequence. Paradoxically, γδ inserts quite randomly throughout target DNAs of heterogeneous or high G+C base composition (10, 43, 44). Studies of insertions in such DNAs showed that the insertion sites are often more A+T-rich than the surrounding DNA (A+T-rich 'valleys'), leading to the suggestion that after γδ lands non-specifically on target DNA, it may 'wander' for a short distance, before inserting into an A+T-rich site (J. Zhang, L. D. Strausbaugh, and C. M. Berg, in preparation).

Tn*9* (IS*1*), which was the first element used for transposon-based DNA sequencing (45, 46), seems to insert with a specificity equivalent to or higher than that of Tn*10*. This has interfered with its usefulness, although a number of Tn*9* derivatives have been constructed for isolating deletions (see Section 4.1).

Phage Mu, which has been used extensively for mutational analysis in *E. coli* and related enteric organisms (3), and also for efficient *in vivo* cloning in enteric bacteria (47, 48), inserts quite randomly in many targets (20), but sometimes exhibits pronounced site-specificity (49). The 5 bp target sequence has a weak G+C-rich consensus sequence (50).

Other elements, including Tn*7* and Tn*554* (from Gram-negative and Gram-positive bacteria, respectively) insert very efficiently into just one highly preferred site per chromosome. Tn*7* transposes with equivalent high efficiency and specificity in several non-enteric organisms, and this has led to its use as a vector for efficient stable placement of specific reporter genes and promoters in a predetermined site (see 3, 25).

2. How to choose and deliver the transposon

Because most transposons move at relatively low frequencies ($\sim 10^{-6}$ to 10^{-4}), a strong and efficient selection scheme is key to their successful use. The choice of transposon should be guided by the goals of the project and whether the target DNA is the host chromosome, a fragment that has already been cloned (and if so, in what vector), or a fragment that has yet to be cloned. For example, if the goal is to obtain one or a few insertions marking a ~100 kb region, insertion hotspots (such as seen with wild-type Tn*10*) are not a major limitation, whereas if an insertion is needed every few hundred base pairs for sequencing, the randomness of insertion is a primary consideration, and use of a member of the Tn*3* family would be advisable. The commonly used elements have been engineered to contain a variety of resistance determinants, reporter genes, promoter genes, and other special features to increase their utility and ease of handling (Table 1) (3).

Strategies for transposon mutagenesis and analysis of chromosomal genes in *E. coli* and *S. typhimurium* have been reviewed recently (see 2, 3, 14) and will not be described in detail here. Strategies for transposon mutagenesis of fragments cloned in phage λ and plasmids, as well as special cloning vectors for isolating transposon-facilitated nested deletions, are described below. Efficient methods for transposon mutagenesis of fragments cloned in the new large-capacity P1 (51) and F-based (BAC) vectors (52, 53) are being developed.

To reduce the size of the transposon, and to increase its stability, most recently constructed transposon derivatives have the element-specific transposition gene(s), for transposase (*tnp*) [and also resolvase (*tnpR*) for replicatively transposing elements] cloned outside of the element. Tn*5* and Tn*10* *tnp* genes work efficiently only *in cis*, so *tnp* is usually engineered into the donor vector, next to a transposon end, while γδ and Tn*3* *tnp* genes work well *in trans*, and are often carried in a separate, compatible, vector. An engineered transposon in a bacterial strain lacking the corresponding transposase does not move because such proteins are highly element-specific.

Tn*5* has been used most widely for mutagenesis of Gram-negative bacteria because it transposes at high frequency and relatively randomly, and its *kan*^{r} gene is

Table 1 Useful transposons and selected derivatives[a]

A. Replicatively transposing elements (Tn*3* family)[e]

Name[b]	Marker(s)	Size (kb)[c]	Special properties[d]
Tn*3*/Tn*1*	*amp*	5	IR, 38 bp. Dupl., 5 bp. Transposition: high frequency to plasmids; into A + T-rich sites; intermolecular transposition forms co-integrates. Because of transposition immunity, cannot be used for pBR-derived targets (11)
mTn*3-lac*	*amp, 'lac*	4.6	Resolution via P1 *loxP* (acted on by site-specific Cre protein); *lac* reporter for selection of protein fusions; also derivatives with genes selectable in *S. cerevisiae* (75)
mTn*3*Cm	*cam* (*'lac*)	1.6	*Cam* gene expressed in *Neisseria* (102); mTn*3*CmNS contains *Nhe*I and *Spe*I restriction sites and mTn*3*Cm*lac* reporter derivative contains promoterless *lac* gene (103)
mini-Tn*3* UFO	*amp*, f1 *ori*		Contains f1 *ori*. When used with f1 *ori*-containing vector, yields single-stranded phage/DNA sequencing templates (76)
mini-Tn*3*-Km	*kan*	1.8	*Campylobacter kan* gene, also expressed in *E. coli* and in *Helicobacter pylori* (104)
γδ (Tn*1000*)	unknown	6	IR, 35 bp. Dupl., 5 bp. Present in *E. coli* F factor. Transposition: high frequency to plasmids; usually random; into A + T-rich sites (valleys) (10, 11)
mγδ-1	*kan*	1.8	Contains γδ res site; Tn*5 kan* gene expressed in *Streptomyces* (44). Also 1.9 kb derivatives with different (multiplex) primer binding sites (10)
pDUAL-1/ pDelta-1	*amp*, pUC *ori* (*kan, tet, sacB, strA* (deletable)	7.9	'Deletion factory' cosmid cloning vectors for isolating nested deletions by intramolecular γδ transposition (see Fig. 5). Contain *amp* and pUC *ori* between γδ ends, and selectable (*kan, tet*) and counterselectable (*sacB, strA*) markers outside γδ ends. pDUAL-1 (33) and pDelta-1 (85) differ in their restriction sites. Can also be used in intermolecuar transposition (G. Wang and C. M. Berg, in preparation)
pDUAL-3	R1 *ori* (*kan, tet, sacB, strA* deletable)	10	Differs from pDUAL-1 and pDelta-1 in lacking *amp*, containing different restriction sites, and in having a low copy number replication origin that is temperature-inducible (C. M. Berg, G. Wang, X. Xu and D. E. Berg, in preparation)
Tn*1721*	*tet*	11.2	IR, 38 bp. Dupl., 5 bp. Member of Tn*21* subgroup. High frequency transposition to plasmids (77). Numerous smaller derivatives with other resistance markers, outward-facing regulated promoter, and reporter genes (105, 106)
TnMax	*cam* or *erm*; *'pho*		fd vegetative replication origin in transposon. No replication origin in vector, and is therefore a suicide vector in strains lacking fd replication genes. Several versions constructed (65)

Table 1 *(continued)*

B. Conservatively transposing elements (diverse families)

Name[b]	Marker(s)	Size (kb)[c]	Special properties[d]
Tn*5*	*kan, ble, str*[f]	5.7; (IS*50*, 1.5)	IR, 8 or 9 bp, but 19 bp needed for transposition. Dupl., 9 bp. Transposition: conservative; high frequency to plasmids and chromosome; into many sites, some hotspots, the *str* gene is silent in *E. coli* (12, 107)
Tn*5seq1*	*kan*	3.2	Tnp^+. Has outward-facing subterminal T7 and SP6 promoter sites for DNA sequencing and riboprobe synthesis (108)
Tn*5-lac*	*kan*; *'lac*	12	Tnp^+. Reporter, forms operon fusions (109). Also a derivative for selection of *lacZ*$^+$ protein fusions and can be converted to *pho*$^+$ fusion, and vice versa, *in vitro* or *in vivo*. Fusion switching used to study subcellular location of site in protein (110). Also small Tnp^- derivatives that form operon (4.9 kb) or protein (5.1 kb) fusions (111)
Tn*phoA*	*kan*; *'phoA*	7.7 or 3.3	Tnp^+ or Tnp^-. Reporter: *phoA*$^+$ protein fusions used to screen for membrane or exported proteins (88, 111)
mini-Tn*5*	*kan, cam, spc-str,* or *tet*; *lacZ, 'phoA,* or *'xylE*		A set of mini-transposons, with various resistance determinants and some with reporter genes. On a mobilizable R6K-based suicide vector (111)
Tn*5-lux*	*tet*, or *kan 'lux*		Tnp^+ or Tnp^-. Reporter: *luxAB* (*Vibrio*) forms operon fusions (111–113)
Tn*5tac*	*kan*	4.6	Tnp^+. Outward-facing regulated *tac* promoter (IPTG-inducible) for generating conditional mutations (91, 94). Also smaller Tnp^- versions (95)
Tn*5supF*	*supF*	<0.3	Insertions selected by suppression of phage λ or chromosomal nonsense mutations (67, 114). Derivative for improved PCR amplification and sequencing (9, 70)
Tn*5-ori*T_S	*kan*; *ori*T_S; *sacB*	~4	Contains *Rhodobacter* transfer origin for Hfr formation and counterselectable *sacB* gene (115)
Tn*5-rpsL*	*kan, strA*$^+$	7.7	Contains counterselectable *strA*$^+$ (*rpsL*) (streptomycin sensitivity) allele, for plasmid curing in strains with chromosomal Str^r (116)
Tn*5*-1087	*cam, erm*	5.3	Tnp^+, for mutagenizing *Anabaena* (117)
Tn*5 -24*	*amp*, T4 gene 24	7.8	Tnp^+, contains phage T4 gene 24 for selection of insertions in phage T4 mutant in gene 24 (118)
Tn*5*-UT set	*kan*; *bar, mer,* or *ars*		Contain *kan* and non-antibiotic resistance determinants (bialaphos, mercury, or arsenite), bracketed by rare restriction sites. On suicide vector (95, 96)
Tn*5*(pfm1)	*kan, cam*		Contains rare restriction sites (*SwaI, PacI, NotI, SfiI, BlnI, SpeI, XbaI*) for pulsed-field mapping of insertion sites (98)
Tn*10*	*tet*	9.3; (IS*10*, 1.3)	IR, imperfect 23 bp, 70 bp needed for optimal transposition; Dupl., 9 bp. Transposition: conservative; to many sites, most insertions in hotspots (16, 119). A mutant transposase with relaxed target specificity results in more random insertion (15, 28)

Table 1 *(continued)*

Name[b]	**Marker(s)**	**Size (kb)**[c]	**Special properties**[d]
Tn*10*HH	*tet*	7.7	Tnp$^+$. Mutations in IS*10* result in higher transposition frequency. Also a 9.4 kb *tet, kan* derivative (120)
Mini-Tn*10*	*tet* (*'lac*)		Several 2.5–5.4 kb derivatives, some encode *kan* and/or a promoterless *lac* reporter gene for selecting operon fusions (120)
Tn*10*-LOF	*kan*; *bar, mer*, or *ars*		Contain *kan* and non-antibiotic resistance determinants (bialaphos, mercury, or arsenite), bracketed by rare restriction sites. Transposons on suicide vector (95, 96)
C. Bacteriophage Mu			
Mu	none	37.5	IR, 2 bp but recognition region ~100 bp; Dupl., 5 bp. Transposition of infecting phage is conservative; intracellular transposition is replicative. Transposition of wild-type Mu and large derivatives almost exclusively to the chromosome; mini-Mu elements also transpose to plasmids. Transposition to many sites, but significant hotspots are found (20, 121)
MudI1	*amp, 'lac*	37	For selection of operon fusions (1). Many derivatives with different selectable genes and/or reporter genes (3, 20)
MudP,Q	*cam*	36.4	Hybrid phage that transpose like Mu and package like P22 (122)
λp_{lac}^{Mu1}	*lac, imm*λ		Hybrid phage that transpose like Mu and package like λ (123). Additional derivatives described in ref. 3
MudII4042	*cam, ori*	16.7	Contains multicopy replication origin from p15A. Used for *in vivo* cloning (47) and generalized transduction (81). Forms protein fusions. Many derivatives with different replication origins, selectable genes and/or reporter genes (3, 20)

[a] Wild-type elements and some of the most useful recent derivatives are described here. A more extensive compilation, including derivatives of less commonly used transposons, can be found in ref. 3. Additional reporter elements are described in ref. 87 and additional Mu derivatives in ref. 20. Independent derivatives with similar phenotypes are grouped together.
[b] Wild-type transposons in bold type.
[c] For composite elements containing a pair of IS elements, the name and size of the IS element are given.
[d] General properties are given for the wild-type elements: IR, essential inverted repeat length (bp); Dupl., target site duplication. The transposase gene(s) are cloned outside the element in most small engineered derivatives (exceptions are marked Tnp$^+$). The appropriate gene must be *in cis* for efficient Tn*5* or Tn*10* transposition, but can be in another plasmid for γδ or Tn*3* transposition.
[e] Intermolecular transposition yields co-integrates (Fig. 1). The wild-type elements contain a resolution site and a site-specific resolvase for efficient resolution of the co-integrate products of intermolecular transposition. Intramolecular transposition does not involve co-integrate formation or a resolution step (Fig. 2).

expressed in many species. The recently described Tn*10* transposase mutations that relax the specificity of Tn*10* target site recognition (36) should increase the usefulness of this element. γδ and other Tn*3*-related elements such as Tn*1721* transpose poorly to the chromosome but are probably the best elements for plasmid mutagenesis because of the randomness with which they transpose.

2.1 Delivering transposons by intermolecular transposition

As described above, transposition from one DNA molecule to another yields a simple insertion (a 'hop') or a co-integrate, depending on whether the transposition event was conservative or replicative, respectively (Fig. 1). The co-integrate can be taken advantage of for the delivery of replicatively transposing elements to plasmids. Replicatively transposing elements are not used for chromosomal mutagenesis because they move preferentially to plasmids.

Simple insertions of conservatively or replicatively transposing elements are often obtained using a suicide donor molecule that cannot persist in the strain under the conditions used for mutagenesis, and selecting for transposon movement to the target replicon (Fig. 1). Insertion of a conservatively transposing element into the chromosome is generally from a suicide phage or plasmid vector. For insertion into plasmids or phage, the element can be introduced by a different phage or plasmid vector, or from the chromosome. The target phage or plasmid, now carrying a transposon insertion, is isolated by phage growth and infection or by plasmid DNA isolation and retransformation.

The co-integrate intermediate formed by replicative transposition from a conjugative plasmid, such as the *E. coli* F factor, has been used to transfer the donor transposon–target plasmid complex to a recipient cell, where resolution occurs to yield a target plasmid with a single simple insertion (Fig. 3). γδ is present on the wild-type F factor, while Tn*3* and γδ derivatives have been placed in a transposon-free derivative of F, pOX38 (54). In each mating a single co-integrate is transferred to the recipient, and then resolved. Therefore, only one mutagenized plasmid is present in most cells. In the conjugal mating system, the transposon need not carry a selectable marker, since transposition is selected by transfer of the co-integrate molecule. The wild-type γδ does not carry a known selectable marker, but most

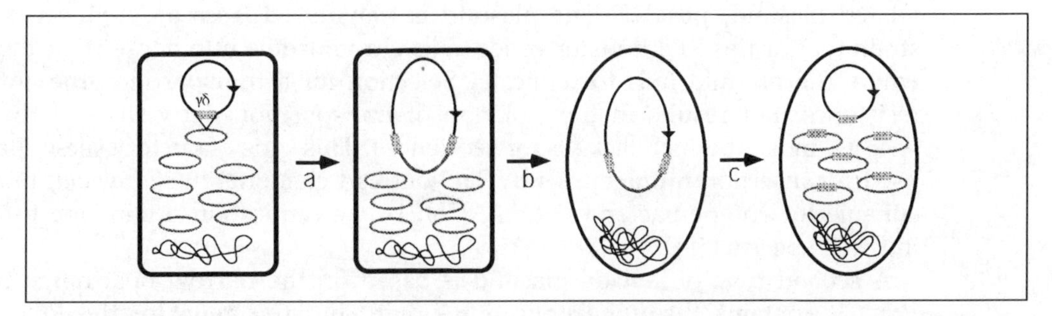

Fig. 3 Selection for insertion into a non-conjugative plasmid by mobilization mediated by transposition of γδ or another member of the Tn*3* family and conjugal transfer of the cointegrate. (a) Transposition from the donor F factor (for wild-type γδ) or pOX38::mini-γδ (for mini-γδ or Tn*3*) to a non-conjugative plasmid, such as pBR322 or a derivative, forming an F::plasmid co-integrate or a pOX38::plasmid cointegrate, as shown in Fig. 1. (b) Conjugation with a plasmid-free F^- recipient strain, selecting for transfer of a plasmid-borne marker (usually ampicillin resistance) under conditions that kill unmated donor and recipient cells. (c) Resolution of the co-integrate in the recipient to yield a plasmid containing one copy of γδ or mini-γδ, and re-forming the donor molecule (F or pOX38::mini-γδ), followed by growth of the transconjugants.

most derivatives do carry one to facilitate subsequent analysis. This ease of delivery and the avoidance of the clone purification step is one reason why co-integrate hopping strategies are useful for large-scale mapping and sequencing of DNAs cloned in plasmids (9, 10, 55).

2.1.1 Suicide vectors

Many clinical or environmental isolates of *E. coli* and most other bacterial species are not sensitive to phages λ or P22. Therefore the phage delivery systems for transposon mutagenesis constructed for use with laboratory strains of *E. coli* K-12 and *S. typhimurium* LT2 are not useful. However, any phage or plasmid vector that can be transferred into a strain but not replicated in it can serve as a suicide vector for transposon delivery. This is illustrated by phage P1, which can transduce, but not replicate in *Myxococcus*, and has proven to be an effective suicide vector for Tn*5* mutagenesis in this organism (56).

A second approach involves broad host range plasmid vectors and superinfection curing by a related incompatible plasmid. For example, in a strategy implemented for Tn*phoA* mutagenesis of *Vibrio cholera*, an *incP* broad host range plasmid containing the transposon is transferred by conjugation to the target strain and selection for the transposon-borne *kan*^{r} determinant. Derivatives in which Tn*phoA* transposition and loss of the vector plasmid occurred are obtained by transfer of a second plasmid of the same compatibility group, but with different drug-resistance markers, and screening for loss of the vector plasmid. Continued growth in the presence of kanamycin, but no selection for either plasmid results in the accumulation of plasmid-free subclones due to spontaneous plasmid curing (57).

Several types of plasmid suicide vectors have been constructed that are suitable for isolating transposon insertions in diverse Gram-negative bacteria. One contains the conjugative DNA transfer site, but not the transfer genes, from a broad host range plasmid, RP4, and the origin of replication from a narrow host range ColE1-related plasmid, pBR322. The plasmid is transferred from an engineered *E. coli* strain that carries RP4 transfer genes in its chromosome into non-enteric recipients where the plasmid fails to replicate. Selection for a transposon-borne antibiotic-resistance trait results in the isolation of transposition derivatives of the target species, most of which lack vector sequences. This type of vector system has been used for insertion mutagenesis by Tn*5*-derived elements in *Rhizobium*, but is not suitable for enteric bacteria because ColE1-type replication origins are functional in these species (58).

A second type of suicide plasmid is based on the narrow host range plasmid R6K. It contains the R6K replication origin, but the gene for the R6K-specific protein has been deleted from the plasmid and placed in the chromosome of the donor strain (59). Once the vector plasmid is transferred to another strain it is separated from the essential protein and fails to replicate. A conjugative derivative also contains the transfer origin of the promiscuous plasmid RP4 (60). When Tn*phoA* (from Tn*5*) was introduced into this plasmid, selection for the transposon-borne marker in a non-permissive host yielded recipients containing transposon insertions. Although this vector system can yield transpositions in a single step, it often yields colonies that retain the vector marker [20% in the case of *Vibrio* matings, but the frequency seems to be species-specific (57)]. Improved versions of such suicide vectors contain transposase genes near to, but outside of, the transposon itself, so that following the initial transposition and loss of the vector plasmid, the element is incapable of moving to new sites (5).

Transferable conditional suicide vectors for Tn*5* have also been generated using a temperature-sensitive derivative of plasmid R388. In such cases, transposition is selected after transferring the plasmid to recipient cells at 30 °C (which allows replication) and a period of growth at 42 °C to block further replication (61). This vector system has also been used with IS*1*-derived transposons to mutagenize *Rhizobium meliloti, Pseudomonas putida, Paracoccus denitrificans, Xanthomonas campestris,* and *Zymomonas mobilis* (62).

A Gram-positive plasmid replication origin, which is inactive in Gram-negative bacteria has been used as a delivery vehicle for a variety of mini-IS*10* derivatives in *E. coli* (63).

Another class of suicide vector contains the single-stranded phage fd replication origin, which functions only in host cells that produce the fd gene 2 protein. Such a suicide plasmid containing Tn*5* has been used for mutagenesis of *E. coli* and *Erwinia amylovora* (64). A set of Tn*1721* (Tn*3* family) derivatives has been constructed on the same principle. However, in these elements the fd origin is between the transposon ends to facilitate subsequent cloning. The transposase and resolvase genes are outside of the transposon. Different versions contain chloramphenicol or erythromycin resistance and a promoterless *phoA* reporter gene (65).

2.1.2 Hopping to phage λ clones (conservative transposition)

The value of simple insertions in selecting for transposon hopping is illustrated by the initial discovery of Tn*5*: λ phage were grown on a bacterial strain that carried an R factor encoding kanamycin resistance. These phage were used to transduce recipient bacteria to kanamycin resistance, and physical characterization of the DNAs of several kanamycin-resistant transducing phage revealed the same ~5.7 kb DNA segment inserted at different sites in λ DNA (66). The selection for kanamycin resistance worked because the λ phage used as a target carried large deletions, and the λ::Tn*5* phage derivatives did not exceed the packaging capacity of the λ phage head.

Because the sizes of many λ clones approaches the limit of what can be packaged in λ, and because most λ cloning vectors were engineered to always grow lytically and kill infected cells, transposons for efficient insertion into λ clones should meet two conditions: the transposon should be small, and it should contain a marker that can be selected during lytic phage growth. A Tn*5* derivative, called Tn*5supF* (67) (Fig. 4a), meets these conditions. The original Tn*5supF* element is only 264 bp long and consists of 19 bp segments from Tn*5* ends and a flanking suppressor tRNA gene, which can be selected during lytic growth.

Tn*5supF* is carried on a plasmid vector, with the Tn*5* transposase gene situated

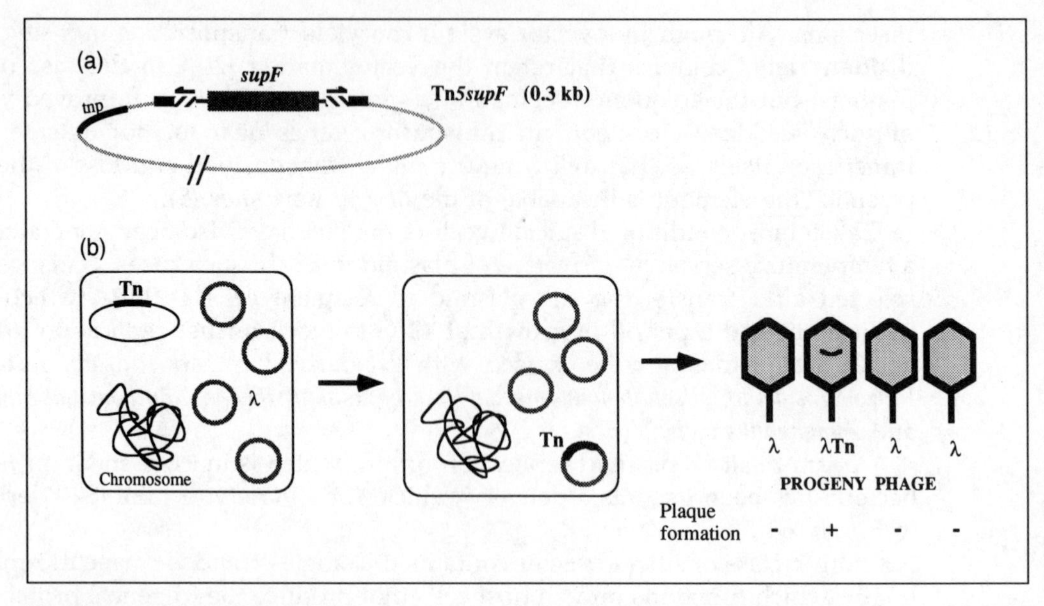

Fig. 4 Transposition of Tn*5supF* to phage λ (see 9, 67, 69, 70, 126). (a) The Tn*5supF* donor plasmid. *supF*, suppressor tRNA gene; *tnp*, Tn*5* transposase gene. (b) Selection for Tn*5supF* insertion into λ by plaque formation on $dnaB_{amber}$ *E. coli* strain DK21.

outside the transposon. Target phage are grown on a Tn*5supF*-containing strain and used to infect $dnaB_{amber}$ cells. Since *dnaB* function is required during λ lytic growth (68), phage carrying Tn*5supF* insertions are easily selected by plaque formation (67, 13) (Fig. 4b). The best of current Tn*5supF* elements contain the efficient M13 forward and reverse primer binding sites, and are carried in a donor plasmid with a highly active Tn*5* transposase gene. They have been useful in large-scale *E. coli* genome sequencing using λ phage clones (9, 70).

2.1.3 Hopping to plasmids (conservative transposition)

Tn*5* and Tn*10* insertions in plasmids are usually obtained by hopping from engineered integration-defective suicide phage vectors (λ in *E. coli* K-12 or P22 in *S. typhimurium*), isolating plasmid DNA, transforming a strain, and selecting for the transposon-borne antibiotic resistance (3). In addition, Tn*5* hops from the chromosome can be selected by increased resistance to kanamycin or neomycin (71). Selection for increased tetracycline resistance cannot, however, be used to select for wild-type Tn*10* hops: when the Tn*10* tet^r determinant is present in multiple copies, the level of resistance actually decreases because excess Tn*10* Tet protein alters the cell membrane (72, 73). Interestingly, the related Tet^r determinant from pBR322 does not show this copy number effect.

The best donor phage vectors are deficient in integrative recombination, and thus unable to insert efficiently into the host chromosome. They also contain amber mutant alleles of phage replication genes, so that lytic replication and cell killing can be minimized by use of non-suppressing bacterial strains (74). The conditional DNA replication mutations are not entirely necessary, however, since phage replication and cell killing can also be blocked by establishing λ *cI* prophage repression. Analogous vector delivery systems have been constructed for *Salmonella*, using derivatives of the *Salmonella* phage, P22.

2.1.4 Hopping to plasmids (replicative transposition)

Historically, the presence of wild-type γδ (Tn*1000*) on the *E. coli* F factor and the ease of obtaining single insertions by conjugal transfer (Fig. 3) led to the wide-spread use of γδ for localizing cloned genes (55), and later for sequencing (3, 10, 42). When the target plasmid is monomeric every transconjugant contains a target plasmid with a single transposon insertion (38). However, wild-type γδ is 6 kb long and does not contain a known marker. This precludes its subsequent use for moving marked alleles to the chromosome (marker exchange). Small ('mini') derivatives of γδ, and of the related elements, Tn*3* and Tn*1721*, that contain the short (<40 bp) terminal inverted repeats, a site for co-integrate resolution, useful restriction sites, primer-binding sites, and an optional selectable marker are described in Table 1.

The wild-type *E. coli* F factor cannot be used to deliver genetically engineered mini-transposon derivatives efficiently because it already contains a copy of wild-type γδ. Therefore, a deletion derivative of F that lacks γδ but remains conjugation-proficient (such as pOX38: see ref. 54), is converted into an F factor analogue by transposition of the mini-transposon from the plasmid in which it was constructed. The resultant pOX38::mini-γδ or pOX38::mini-Tn*3* plasmid is then used, like wild-type F, to deliver the transposon to target plasmids by mating (Fig. 3) (10, 44, 75). Transposase is provided *in trans* from a helper plasmid in the donor strain, and resolvase is provided *in trans* from a helper plasmid in the recipient strain, or from copies of γδ resident in the chromosome.

The original F^+ *E. coli* K-12 isolate contained γδ in the F factor, but apparently not in its chromosome. During cultivation in the laboratory, many F^- strains ended up with one or more copies of γδ in the chromosome (54). These strains do not produce enough transposase to catalyse efficient transposition, but they do produce enough resolvase to catalyse efficient resolution of co-integrates. Therefore if a $γδ^+$ strain is used as recipient, a cloned resolvase gene is not needed. It is preferable to maintain mini-γδ in a strain that lacks γδ, when possible. γδ-free strains include AB1157, C600, CBK884, HB101, JM109, MG1655, RR1, and W3110. γδ-containing strains include DH5α, DH10B, and MG1043 (54; F. LaBanca, X. Xu, G. Wang, and C. M. Berg, unpublished). We find that DH10B (or CBK940, a Nal^r derivative of DH10B) is a good recipient for mini-γδ matings because it contains γδ in the chromosome and yields high quality DNA preparations for sequencing.

It should be noted that, after mating, the recipient cell contains copies of γδ in both the donor and target replicons. None the less, when a multicopy target plasmid is used, good quality templates for sequencing with γδ primers are obtained despite the presence of trace amounts of contaminating donor plasmid (42).

Although the first detailed analysis of transposition by the closely related ampicillin resistance (amp) element, Tn*3*, indicated that the Tn*3* moves very non-randomly (37), these results appear to be atypical: Tn*3* has been used successfully for mutagenesis and sequencing of other plasmids (76). However, use of Tn*3* requires special vectors (75) because most commonly used pBR- or pUC-related plasmids retain one end of Tn*3*, and Tn*3*-specific transposition immunity severely reduces transposition frequency (11, 22, 77). Since γδ and Tn*3* are closely related, it is likely that they are equivalent in transposition specificity.

In the γδ/Tn*3* mating delivery strategy it is important that the target plasmid be maintained as a monomer. If it is a dimer, or present as a mixture of monomers and dimers, as is common in Rec^+ strains (78), recovery of insertion-containing plasmids is inefficient. This is because a transposon-containing dimer will have one γδ-free component that can segregate from its γδ-containing sibling after recombination and may be selected during bacterial growth (38). Therefore, a Rec^- donor strain should be used whenever possible.

An elegant set of Tn*1721* (Tn*21* subgroup of Tn*3* family: 77) derivatives has been constructed for use in a suicide, rather than a mating strategy. They contain the phage fd origin between the transposon ends, with the transposase and resolvase genes outside the transposon. When introduced into a strain that lacks the fd phage replication protein, encoded by gene 2, selection for the transposon marker yields insertions into the target plasmid, immediately followed by resolution and loss of the donor plasmid. Different versions of these transposons, called TnMax, contain chloramphenicol or erythromycin resistance markers and a promoterless *phoA* reporter gene (65).

2.1.5 Phage Mu

Phage Mu is a versatile, broad host-range transposable element. It earned its name (mutator phage) because a high frequency of mutations were found among newly formed lysogens (79), due to Mu phage insertion into many genomic sites, sometimes in genes that affect phenotypic traits. Lysogenization of Mu phage occurs by conservative transposition. Induction of a Mu lysogen, on the other hand, results in many rounds of replicative transposition (Mu replicates only by transposing!), associated with extensive genomic rearrangement (multiple rounds of the transposition events depicted in Fig. 1, right and in Fig. 2B). Copies of Mu are inserted at many sites, often within a few kilobases of each other. At the end of the lytic cycle, packaging is initiated just outside of the Mu left (*c*) end and approximately 39 kb of DNA are incorporated into each phage head. The wild-type Mu genome is 37.5 kb, and thus about 1.5 kb of bacterial DNA adjacent to the Mu right (*S*) end is also packaged by a 'headful' mechanism. Deletion derivatives, called mini-Mu, package correspondingly larger amounts of host DNA. This is important for mini-Mu-mediated *in vivo* cloning (see below). Since Mu has a broad host range, the delivery vehicle is usually Mu itself. However, Mu can also be delivered by conjugation or by P1 transduction (20).

Mu was the first transposable element to be genetically engineered (to contain an ampicillin resistance marker and a promoterless *lac* reporter gene) (1). This and other Mu derivatives have been widely used to isolate mutations in chromosomal genes, and to study their regulation. The *lac* reporter gene has been used in the identification and characterization of many previously unknown genes that confer no obvious phenotype, such as genes involved in repairing DNA damage or the heat shock response (3, 80). Over a hundred Mu and mini-Mu derivatives have been constructed (20) (Table 1).

Wild-type Mu inserts poorly into plasmids, but small, mini-Mu derivatives can be used for insertion into plasmids (20, 49). The widespread belief that Mu transposes randomly is belied by the many hotspots found in plasmids (49), and probably also by clustered insertions into the chromosome (see ref. 3).

Mini-Mu elements are also useful for generalized transduction. Most mini-Mu transducing phage, unlike P1 transducing phage, contain both phage and bacterial DNA: they arise by headful packaging of phage genomes that are too small to fill the head, and therefore incorporate adjacent host DNA. The amount of host DNA carried by mini-Mu variants ranges up to only about 32 kb [the difference between the minimal element size and the Mu packaging capacity (39 kb)], in contrast to the ~100 kb carried by the generalized transducing phage P1. This is useful in localized mutagenesis, in which mutations clustered close to a selectable gene are isolated by selecting those that are cotransduced from a large random pool. The clustering is tighter using mini-Mu lysates than is possible with phage P1 (81).

3. *In vivo* cloning: Mu and D3112

The replicative transposition and headful packaging properties of phage Mu were used to construct a series of mini-Mu elements for *in vivo* cloning (47, 48). These elements contain the essential Mu ends, a selectable gene, and a plasmid replication origin. Some of the products formed during replicative transposition (lytic growth) (analogous to that depicted in Fig. 1, right) have two mini-Mu elements close enough to be packaged in the same phage head. After infection of a Rec^+ strain, homologous recombination between two Mu DNAs that had been inserted in the same orientation, yields a circular molecule capable of self-replication, a mini-Mu plasmid.

Since mini-Mu cloning sites are determined by Mu insertion, not by restriction sites, a set of clones can also be used to localize the selected gene to the region of overlap (82). Use of one of the mini-Mus that contains a *lac* reporter gene also permits the identification of adjacent genes (48, 82).

D3112, a *Pseudomonas aeruginosa* phage related to Mu, has also been engineered for *in vivo* cloning in this genus (83).

4. Cosmid cloning vectors for delivering transposon ends in intramolecular transposition: nested deletion formation and DNA sequencing

As described above, one product of intramolecular transposition is a deletion plasmid with one endpoint at the transposon end and the other at the insertion site in the plasmid (Fig. 2). These deletion products are very useful for molecular and genetic analyses, because the deletion endpoint can be deduced directly from plasmid size, and because endpoints at particular sites can be selected from large pools of random deletion plasmids by agarose gel electrophoresis (9). Deletions are rare, but can easily be isolated in plasmids that contain a counterselectable (conditional lethal) gene adjacent to the transposon (9, 27, 33, 84, 85, 86).

The *in vivo* transposon-based nested deletion method is better than *in vitro* nested deletion methods for analysing large cloned DNA segments because the deletions are easy to isolate in large as well as small cloned DNA, and can extend over many kilobases. A nested deletion strategy is particularly attractive when the DNA of interest is too large for easy (e.g. PCR-based) mapping of simple insertions. Gaps in an array of deletions are easily located and can be filled by primer walking or by size fractionation of a pool of deletion plasmids.

4.1 Vectors for accessing one target strand

The very first implementation of transposon-based sequencing used Tn*9*/IS*1* in a nested deletion strategy (84, 86). These vectors contain a counterselectable (suicide) gene (*galKT*$^+$ in a *galE* strain) between the transposon and the cloning site. Intramolecular transposition generates deletions that extend into the cloned DNA in either direction (see Fig. 2B). Deletions that lose *galKT*$^+$ and retain the essential plasmid replication origin and antibiotic resistance markers are viable. Those that have a deletion endpoint in the cloned fragment are useful for sequencing.

This system has two serious disadvantages: firstly, IS*1* transposes with high specificity, which makes it difficult to obtain complete coverage of some DNA segments; and, secondly, in order to access both strands for DNA sequencing, the fragment must be cloned in both orientations. As a result, few groups are currently using IS*1*-based nested deletion strategies.

4.2 Vectors for accessing both target strands: pDUAL/pDelta

Both disadvantages of the IS*1*-based vectors noted above have been addressed in the pDUAL/pDelta vector series: γδ, which transposes more randomly than IS*1*, was used, and both target strands can be accessed in a single vector. To access both strands, clockwise and counterclockwise deletion products must be viable and also individually selectable. To accomplish this, the plasmid replication origin and selectable marker(s) must be present either in, or on each side of, the transposon so that no unique essential information is located outside the transposon.

The pDUAL γδ-based plasmids contain a replication origin between the transposon ends, and appropriate pairs of counterselectable and selectable genes outside of the element. The *sacB*$^+$ and *strA*$^+$ counterselectable genes on each side of γδ allow nested deletions extending into the cloned fragment in either direction to be selected easily using simple media containing sucrose or streptomycin, respectively (33, 85). The selectable *kan* and *tet* genes (encoding kanamycin and tetracycline resistance, respectively) between *sacB* and *strA* and the cloning site are used to eliminate deletions that extend beyond the ends of the cloned fragment. This selection *against* one counterselectable marker and *for* one selectable marker yields colonies with plasmids whose nested deletions extend in the desired direction for varying distances into the cloned target DNA. After transposition, one transposon end always abuts a deletion endpoint, and the unique sequence just inside the transposon end can serve as a 'universal' primer binding site for sequencing adjacent cloned DNA. All target regions can be sequenced using a pair of 'universal' primer binding sites engineered into γδ. Outward-facing SP6 and T7 promoters just within the γδ ends serve as sites for universal primer binding and riboprobe synthesis after γδ-mediated deletion formation (Fig. 5). pDUAL-1 (33) and pDelta-1 (85) contain the pUC19 high copy number replication origin and *amp* gene, but different unique restriction sites (9). pDUAL-3 contains an amplifiable low copy number replication origin to allow greater stability of segments not clonable in pUC-based vectors (C. M. Berg, G. Wang, X. Xu, and D. E. Berg, unpublished).

5. Special uses of transposons

Here we emphasize some particularly important engineered transposons and their uses.

5.1 As mobile reporter elements

Transposons can be used as reporter elements for analysing transcription, translation, and the location of gene products. Such uses of gene fusions have revolutionized the analysis of gene regulation, and led to the discovery of many new genes, including ones that respond to environmental signals such as heat shock and DNA damage.

As noted above, the first engineered transposon was a phage Mu derivative with a promoterless *lac* gene at one end (1). This element, Mud1, introduced the concept of engineering reporter genes in transposons. There are two classes of reporter genes: 'operon fusion' genes, like a *lac* in the original Mud1, that contain a translational start signal but not transcriptional start signals, and 'protein fusion' genes that do not contain either kind of start signals and make hybrid proteins when inserted into a gene in the correct orientation and correct reading frame. Different reporter elements are described in Table 1 and more extensively in ref. 87.

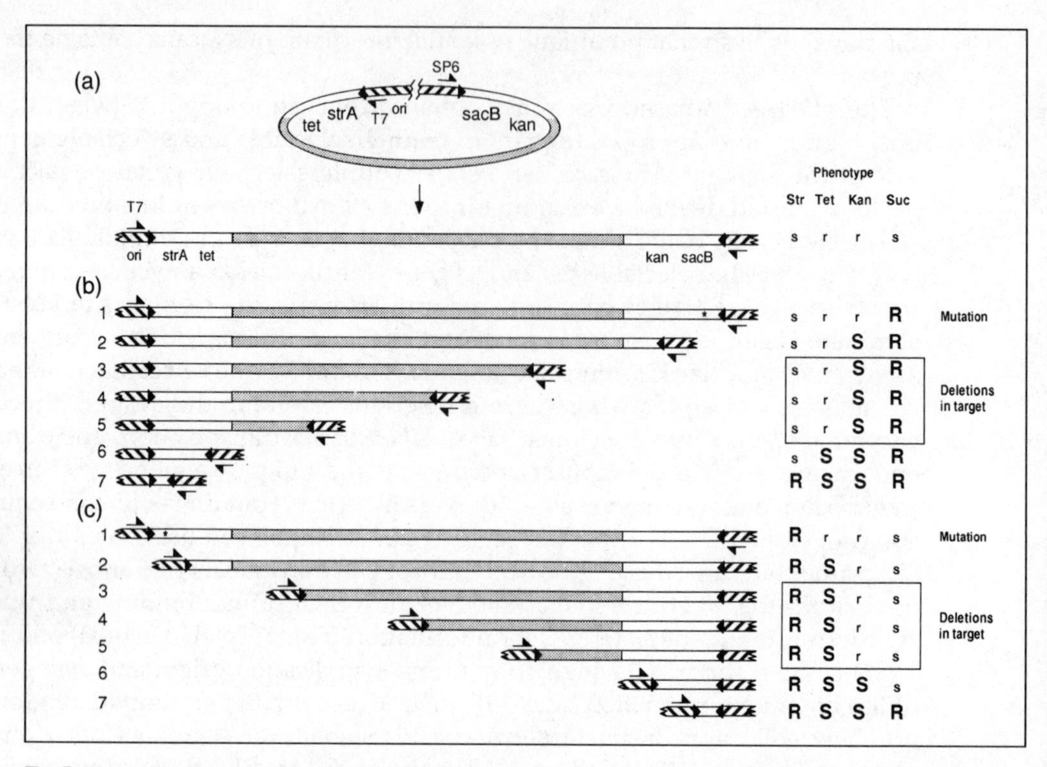

Fig. 5 Expected phenotypes of pDUAL deletion derivatives. (a) Linearized parent plasmid. (b) Plasmids from colonies selected in sucrose plates: 1, mutation in *sacB* (or very small deletion that does not extend into *kan*); 2, deletion extending into *kan*, but not into the target DNA; 3–5, deletions extending various distances into the cloned fragment; 6 and 7, deletions extending beyond the cloned fragment. (c) Plasmids from colonies selected in streptomycin plates: as in panel (b) but *strA* substituted for *sacB*. Deletions 6 and 7 would not be recovered because the selection plates contain tetracycline (in b) or kanamycin (in c). Selection is for Suc^r, Tet^r (Kan^S) versus Str^r, Kan^r (Tet^S) colonies. The only plasmids recovered that do not have deletion endpoints in the cloned fragment are those with deletion endpoints in *kan* or *tet*, respectively (relatively few if the cloned fragment is large). pDUAL 1 (33) and pDelta-1 (85) contain the replication origin and *amp* from pUC19 between the γδ ends, but differ in their restriction sites. pDUAL-3 (C. M. Berg *et al.*, unpublished) contains an amplifiable R1 replication origin and no *amp* gene.

Most reporter transposons contain *lac* as the indicator gene because β-galactosidase is easy to assay visually or enzymatically. β-galactosidase functions only when it is present in the cytoplasm. A number of other reporter genes have been used to detect genes encoding membrane-bound and periplasmic proteins. Some are more readily selectable than *lac*, and allow selection in eukaryotes. These include *cam* (which encodes chloramphenicol transacetylase), *kan* (aminoglycoside phosphotransferase), *lux* (luciferase), *phoA* (alkaline phophatase), and *uid* (β-glucuronidase) (Table 1) (6, 87). The most important of these is *phoA*.

To map the domains of proteins that traverse the cell membrane, and to monitor easily the regulation of expression of genes encoding such proteins, a *phoA*–alkaline phosphatase protein fusion reporter derivative of Tn*5* called Tn*phoA* was constructed (88). The utility of the Tn*phoA* transposon is based on earlier findings that alkaline phosphatase is activated by passage through a bacterial cell membrane, and that such passage depends on an N-terminal signal sequence. When inserted in the correct reading frame, the *phoA* gene near an end of Tn*phoA* forms protein fusions that lack the *phoA* signal sequence. Mutant proteins remain intracellular and inactive unless *phoA* is fused with another gene whose product contains an N-terminal signal sequence. Tn*phoA* can be delivered from a λ phage vector, from an F factor plasmid, or from a transferable suicide vector that functions in enteric as well as non-enteric organisms. In addition to its original use for probing protein conformation, Tn*phoA* has been used to identify genes likely to be important in virulence since many proteins involved in bacterial pathogenesis (e.g. those involved in motility, adherence, and toxicity) are surface or secreted proteins.

A derivative of Tn*phoA* called Tn*phoA*-IG that contains the sites used for single strand DNA synthesis and packaging into f1 particles (the 'intergenic region') was constructed to facilitate the sequencing of *phoA* insertion sites in genes cloned in small plasmids. It can be delivered from an F factor single copy vector, and insertions in multicopy plasmids selected by their hyper-resistance to kanamycin, and packaging into f1 particles after helper phage infection (89).

A Mud*phoA* transposon, equivalent to Tn*phoA*, was constructed especially for use in species such as *Legionella pneumophila*, in which Tn*5* transposition is rare and Mu transposition is quite efficient (90). Exchange of the reporter sequences between Tn*phoA* and Tn*lac* insertions (fusion switching) permits high resolution analysis of protein localization (110).

5.2 As mobile promoter elements

Transposons can also provide mobile promoters to turn on, or turn off, adjacent genes. A Tn*5* derivative called Tn*5tacl* allows isolation of insertion mutations with conditional mutant phenotypes. This element contains an outward facing promoter, which is regulated by the *lac* repressor gene, *lacI*, a kanamycin resistance gene, and the Tn*5* transposase gene (91). Tn*5tac1* can be delivered to the *E. coli* genome from a λ phage vector (91), and to non-enteric bacteria from a suicide plasmid vector (92). Tests with representative Tn*5tac1* insertions showed that mutant phenotypes result from IPTG-induced excessive transcription of a gene or site, in cases of insertions upstream of the target gene (91), and also from IPTG-induced decreases in expression, in cases of insertion downstream of the target gene (93, 94).

A mini-Tn*5* derivative containing an outward facing, *lacI*-regulated promoter (P_{trc}) was described recently (95). This transposon is in the pUT broad host range conjugal suicide vector (96) that is useful for mutagenizing enteric and non-enteric Gram-negative bacteria. The vector origin of replication requires products of replication genes implanted in the chromosome of the *E. coli* donor strain. The transposase gene is also in the vector, but outside of the transposon.

5.3 In genome mapping and sequencing studies

The earliest uses of transposons were to isolate insertions in or adjacent to chromosomal genes and to map the mutations genetically by inheritance of the associated antibiotic resistance marker. Insertions in hundreds of sites in the *E. coli* chromosome have been described and mapped (2, 4). A very useful set of insertions are marked by kanamycin or tetracycline resistance at about 1 min intervals along the *E. coli* chromosome (4).

Insertions have also been used to clone genes by digesting chromosomal DNA of insert-containing strains, ligating into a standard cloning vector (or, in the case of transposons that carry a replication origin, self-ligating) and selecting for the drug-resistance determinant. Cloning the marked mutant allele is particularly valuable when the wild-type allele is difficult to select for, or lethal in multiple copies. Once the insertion-containing allele has been cloned and analysed, if a clone containing the functional allele is viable, it can be isolated or reconstituted by a number of strategies including marker exchange.

Pulsed field gel electrophoresis of genomic DNA cut by a restriction enzyme that yields few large restriction fragments per genome has become a valuable technique for genome analysis. Usually, genes are mapped to these fragments by hybridization, but McClelland and colleagues have exploited rare restriction sites present in wild-type Tn*5* (97) and engineered genomic DNA into a mini-Tn*5* derivative (98), to map genes to fragments directly by digestion with a restriction enzyme that cuts the transposon and determining the band that is lost.

On a smaller scale, transposons can be used to good advantage to reduce the labour involved in constructing a fine-structure genomic restriction map, such as that constructed for *E. coli* (99). The positions of insertions in large plasmid or cosmid clones can be mapped by hybridization (10, 43) and then used to order the smaller restriction fragments in that clone by change in size. Another promising approach for generating restriction maps of DNAs cloned in the pDUAL/pDelta vectors described above, involves ordering restriction fragments by their 'drop-out' patterns in deletion clones (C. M. Berg *et al.*, unpublished).

As discussed above, any transposon insertion can be used to provide a mobile primer binding site for DNA sequencing. Therefore, insertions isolated for gene identification, mapping, expression studies, or other purposes, can be used to obtain the sequence of the adjacent DNA, whether it is cloned or in the genome. Transposon-based methods also merit consideration as adjuncts to other sequencing methods. For example, although primer walking is currently used primarily to fill gaps because of the cost of making many primers that are used only once, this approach will become more economical as walking with preformed primers (100) becomes feasible. Insertions can be used with primer walking to provide multiple access sites in a large plasmid or cosmid clone. This will be particularly important for sequencing tandem repeats that are too homogeneous to be aligned by shotgun methods and too long to be bridged by a long sequence run (9, 10).

With the great variety of transposons available, the choice of the element and approach to be used should be determined by the overall goals of the project, what is already known about the organism, and the materials available. At one extreme, if the goal is to construct a genomic restriction map and to sequence selected portions of the genome of an organism that has not been studied previously, cloning into a pDUAL cosmid cloning vector (9, 33, 85) would be advantageous. At the other extreme, the entire *E. coli* genome has been cloned, and more than half sequenced to date (101). A gene from a region that has not yet been sequenced can usually be cloned 'by mail', and accessed most easily by Tn*5supF* (for λ clones) or mini-γδ (for plasmid or cosmid clones) (9, 10).

6. Future prospects

Transposable elements have been used to facilitate studies of gene identification, mapping, regulation, and sequencing in *E. coli* and related species since the discovery of phage Mu and IS elements in the 1960s and of transposons in the 1970s. Improved transposons and transposon-based vectors are still being constructed and tested for many specialized purposes. We anticipate that transposons will continue to be valuable for molecular genetic analyses in an expanding number of organisms.

Acknowledgements

We thank our former and present colleagues and friends for their many contributions to the work and ideas presented here. We are particularly indebted to Robert Blakesley, Henry Huang, Katsumi Isono, Frank LaBanca, Lin Liu, David Sherratt, Linda Strausbaugh, Gan Wang, and Xiaoxin Xu. Work in our laboratories was supported by grants from the Department of Energy (DE-FG02-89ER-60862 and DE-FG02-90ER-610), the National Institutes of Health (HG-000563), Life Technologies Incorporated, and the University of Connecticut Research Foundation.

References

1. Casadaban, M. J. and Cohen, S. N. (1979) Lactose genes fused to exogenous promoters in one step using a Mu-lac bacteriophage: *in vivo* probe for transcriptional control sequences. *Proc. Natl Acad. Sci. USA*, **76**, 4530.
2. Berg, C. M. and Berg, D. E. (1987) Uses of transposable elements and maps of known insertions. In *Escherichia coli and Salmonella typhimurium: Cellular and Molecular Biology*. F. C. Neidhardt, J. L. Ingraham, K. B. Low, B. Magasanik, M. Schaechter, and H. E. Umbarger (eds). American Society for Microbiology, Washington DC, p. 1071.
3. Berg, C. M., Berg, D. E., and Groisman, E. A. (1989) Transposable elements and the genetic engineering of bacteria. In *Mobile DNA*. D. E. Berg and M. M. Howe (eds). American Society for Microbiology, Washington, DC, p. 879.
4. Singer, M., Baker, T. A., Schnitzler, G., Deischel, S. M., Goel, M., Dove, W., Jaacks, K. J., Grossman, A. D., Erickson, J. W., and Gross, C. A. (1989) A collection of strains containing genetically linked alternating antibiotic resistance elements for genetic mapping of *Escherichia coli*. *Microbiol. Rev.*, **53**, 1.
5. de Lorenzo, V. and Timmis, K. N. (1994) Analysis and construction of stable pheno-

types in Gram-negative bacteria with Tn*5* and Tn*10*-derived mini-transposons. *Methods Enzymol.*, **235,** 386–405.

6. Miller, J. H. (1992) *A short course in bacterial genetics—a Laboratory Manual and Handbook for Escherichia coli and Related Bacteria*. Cold Spring Harbor Laboratory Press, Plainview, NY.
7. Berg, D. E. and Howe, M. M. (eds) (1989) *Mobile DNA*. American Society for Microbiology, Washington, DC.
8. Miller, J. H. (ed.) (1991) Bacterial genetic systems. *Methods Enzymol.*, Volume **204**. Academic Press, New York.
9. Berg, C. M., Wang, G., Isono, K., Kasai, H., and Berg, D. E. (1994) Transposon-facilitated large-scale DNA sequencing. In *Automated DNA Sequencing and Analysis Techniques*. M. D. Adams, C. Fields and C. Venter (eds). Academic Press, London, p. 51.
10. Berg, C. M., Wang, G., Strausbaugh, L. D., and Berg, D. E. (1993) Transposon-facilitated sequencing of DNAs cloned in plasmids. *Methods Enzymol.*, **218,** 279.
11. Sherratt, D. (1989) Tn*3* and related transposable elements: site-specific recombination and transposition. In *Mobile DNA*. D. E. Berg and M. M. Howe (eds). American Society for Microbiology, Washington, DC, p. 163.
12. Berg, D. E. (1989) Transposon Tn*5*. In *Mobile DNA*. D. E. Berg, and M. M. Howe (eds). American Society for Microbiology, Washington, DC, p. 185.
13. Krishnan, B. R., Kersulyte, D., Brikun, I., Huang, H. V., Berg, C. M., and Berg, D. E. (1993) Transposon-based and polymerase chain reaction-based sequencing of DNAs cloned into lambda phage. *Methods Enzymol.*, **218,** 258.
14. de Bruijn, F. J. (1987) Transposon Tn*5* mutagenesis to map genes. *Methods Enzymol.*, **154,** 175.
15. Kleckner, N., Bender, J., and Gottesman, S. (1991) Uses of transposons with emphasis on Tn*10*. *Methods Enzymol.*, **204,** 139.
16. Kleckner, N. (1989) Transposon Tn*10*. In *Mobile DNA*. D. E. Berg and M. M. Howe (eds). American Society for Microbiology, Washington, DC, p. 227.
17. Caparon, M. G. and Scott, J. R. (1991) Genetic manipulation of pathogenic Streptococci. *Methods Enzymol.*, **204,** 556.
18. Youngman, P. (1987) Plasmid vectors for recovering and exploiting Tn*917* transpositions in *Bacillus* and other Gram-positive bacteria. In *Plasmids: A Practical Approach*. K. G. Hardy (ed.). IRL Press, Oxford, p. 79.
19. Symonds, N., Toussaint, A., van de Putte, P., and Howe, M. M. (eds) (1987) *Phage Mu*. Cold Spring Harbor Laboratory Press, Cold Spring Harbor, NY.
20. Groisman, E. (1991) *In vivo* genetic engineering with bacteriophage Mu. *Methods Enzymol.*, **204,** 180.
21. Mizuuchi, K. (1992) Transpositional recombination: mechanistic insights from studies of mu and other elements. *Annu. Rev. Biochem.*, **61,** 1011.
22. Heffron, F. (1983) Tn*3* and its relatives. In *Mobile genetic elements*. J. A. Shapiro (ed.). Academic Press, NY. p. 223.
23. Bishop, R. and Sherratt, D. (1984) Transposon Tn*1* intra-molecular transposition. *Mol. Gen. Genet.*, **196,** 117.
24. Wang, G., Xu, X., Chen, J., Berg, D. E., and Berg, C. M. (1994) Inversions and deletions generated by a mini-γδ (Tn*1000*) transposon. *J. Bacteriol.*, **176,** 1332.
25. Craig, N. L. (1989) Transposon Tn*7*. In *Mobile DNA*. D. E. Berg and M. M. Howe (eds). American Society for Microbiology, Washington, DC, p. 211.
26. Biel, S. W. and Berg, D. E. (1984) Mechanism of IS1 transposition in *E. coli:* choice between simple insertion and cointegration. *Genetics*, **108,** 319.

27. Tomcsanyi, T., Berg, C. M., Phadnis, S. H., and Berg, D. E. (1990) Intramolecular transposition by a synthetic IS*50* (Tn*5*) derivative. *J. Bacteriol.*, **172,** 6348.
28. Bender, J. and Kleckner, N. (1992) Tn*10* insertion specificity is strongly dependent upon sequences immediately adjacent to the target-site consensus sequence. *Proc. Natl Acad. Sci. USA*, **89,** 7996.
29. Weinert, T. A., Schaus, N. A., and Grindley, N. D. F. (1983) Insertion sequence duplication in transpositional recombination. *Science*, **222,** 755.
30. Tsai, M.-M., Wong, R. Y.-P., Hoang, A. T., and Deonier, R. C. (1987) Transposition of Tn*1000*: *in vivo* properties. *J. Bacteriol.*, **169,** 5556.
31. Foster, T. J. (1975) Tetracycline-sensitive mutants of the F-like factors R100 and R100-1. *Mol. Gen. Genet.*, **137,** 85.
32. Bochner, B. R., Huang, H.-C., Schieven, G. L., and Ames, B. N. (1980) Positive selection for loss of tetracycline resistance. *J. Bacteriol.*, **143,** 926.
33. Wang, G., Blakesley, R. W., Berg, D. E., and Berg, C. M. (1993) pDUAL: a transposon-based cosmid cloning vector for generating nested deletions and DNA sequencing templates *in vivo*. *Proc. Natl Acad. Sci. USA*, **90,** 7874.
34. Chandler, M., Roulet, E., Silver, L., Boy de la Tour, E., and Caro, L. (1979) Tn*10* mediated integration of the plasmid R100.1 into the bacterial chromosome: inverse transposition. *Mol. Gen. Genet.*, **173,** 23.
35. Nag, D. K., DasGupta, U., Adelt, G., and Berg, D. E. (1985) IS*50*-mediated inverse transposition: specificity and precision. *Gene*, **34,** 17.
36. Bender, J. and Kleckner, N. (1992) IS*10* transposase mutations that specifically alter target site recognition. *EMBO J.*, **11,** 741.
37. Tu, C. P. and Cohen, S. N. (1980) Translocation specificity of the Tn*3* element: characterization of sites of multiple insertions. *Cell*, **19,** 151.
38. Liu, L. and Berg, C. M. (1990) Mutagenesis of dimeric plasmids by the transposon γδ (Tn*1000*). *J. Bacteriol.*, **172,** 2814.
39. Wiater, L. A. and Grindley, N. D. F. (1990) Integration host factor increases the transpositional immunity conferred by γδ ends. *J. Bacteriol.*, **172,** 4951.
40. Schwacha, A., Cohen, J. A., Gehring, K. B., and Bender, R. A. (1990) Tn*1000*-mediated insertion mutagenesis of the histidine utilization (hut) gene cluster from *Klebsiella aerogenes*: genetic analysis of *hut* and unusual target specificity of Tn*1000*. *J. Bacteriol.*, **172,** 5991.
41. Reed, R. R., Young, R. A., Steitz, J. A., Grindley, N. D. F., and Guyer, M. S. (1979) Transposition of the *Escherichia coli* insertion element γ generates a five-base-pair repeat. *Proc. Natl Acad. Sci. USA*, **76,** 4882.
42. Liu, L., Whalen, W., Das, A., and Berg, C. M. (1987) Rapid sequencing of cloned DNA using a transposon for bidirectional priming: sequence of the *Escherichia coli* K-12 *avtA* gene. *Nucleic Acid Res.*, **15,** 9461.
43. Strausbaugh, L. D., Bourke, M. T., Sommer, M. T., Coon, M. E., and Berg, C. M. (1990) Probe mapping to facilitate transposon-based DNA sequencing. *Proc. Natl Acad. Sci. USA*, **87,** 6213.
44. Berg, C. M., Vartak, N. B., Wang, G., Xu, X., Liu, L., MacNeil, D. J., Gewain, K. M., Wiater, L. A., and Berg, D. E. (1992) The mγδ-1 element, a small γδ (Tn*1000*) derivative useful for plasmid mutagenesis, allele replacement and DNA sequencing. *Gene*, **113,** 9.
45. Ahmed, A. (1984) Use of transposon-promoted deletions in DNA sequence analysis. *J. Mol. Biol.*, **178,** 941.

46. Ahmed, A. (1987) Use of transposon-promoted deletions in DNA sequence analysis. *Methods Enzymol.*, **155,** 177.
47. Groisman, E. A., Castilho, B. A., and Casadaban, M. J. (1984) *In vivo* DNA cloning and adjacent gene fusing with a mini-Mu–lac bacteriophage containing a plasmid replicon. *Proc. Natl Acad. Sci. USA,* **81,** 1480.
48. Groisman, E. A. and Casadaban, M. J. (1987) Cloning of genes from members of the family *Enterobacteriaceae* with mini-Mu bacteriophage containing plasmid replicons. *J. Bacteriol.*, **169,** 687.
49. Castilho, B. A. and Casadaban, M. J. (1991) Specificity of mini-Mu bacteriophage insertions in a small plasmid. *J. Bacteriol.*, **173,** 1339.
50. Mizuuchi, K. (1983) *In vitro* transposition of bacteriophage Mu: a biochemical approach to a novel replication reaction. *Cell,* **35,** 785.
51. Sternberg, N. (1990) A bacteriophage P1 cloning system for the isolation, amplification and recovery of DNA fragments as large as 100 kb. *Proc. Natl Acad. Sci. USA,* **87,** 103.
52. Hosada, F., Nishimura, S., Uchida, H., and Ohki, M. (1990) An F factor based cloning system for large DNA fragments. *Nucleic Acids Res.*, **18,** 3863.
53. Shizuya, H., Birren, B., Kim, U.-J., Mancino, V., Slepak, T., Tachiiri, Y., and Simon, M. (1992) Cloning and stable maintenance of 300-kilobase-pair fragments of human DNA in *Escherichia coli* using an F-factor-based vector. *Proc. Natl Acad. Sci. USA,* **89,** 8794.
54. Guyer, M. S., Reed, R. R., Steitz, J. A., and Low, K. B. (1980) Identification of a sex-factor-affinity site in *E. coli* as gamma-delta. *Cold Spring Harbor Symp. Quant. Biol.*, **45,** 135.
55. Guyer, M. S. (1983) Uses of the transposon $\gamma\delta$ in the analysis of cloned genes. *Methods Enzymol.*, **101,** 362.
56. Kuner, J. M. and Kaiser, D. (1981). Introduction of transposon Tn*5* into *Myxococcus* for analysis of developmental and other nonselectable mutants. *Proc. Natl Acad. Sci. USA,* **78,** 425.
57. Taylor, R. K., Manoil, C., and Mekalanos, J. J. (1989) Broad host range vectors for delivery of Tn*phoA*: use in genetic analysis of secreted virulence determinants of *Vibrio cholerae. J. Bacteriol.*, **171,** 1870.
58. Simon, R., Priefer, U., and Puhler, A. (1983) Vector plasmids for *in vivo* and *in vitro* manipulations of Gram-negative bacteria. In *Molecular Genetics of the Bacteria–Plant Interactions*. A. Puhler (ed.) p. 98. Springer Verlag, Berlin.
59. Kolter, R., Inuzuka, M., and Helinski, D. R. (1978) Trans-complementation-dependent replication of a low molecular weight origin fragment from plasmid R6K. *Cell,* **15,** 1199.
60. Miller, V. L. and Mekalanos, J. J. (1988) A novel suicide vector and its use in construction of insertion mutations: osmoregulation of outer membrane proteins and virulence determinants in *Vibrio cholerae* requires *toxR*. *J. Bacteriol.*, **170,** 2575.
61. Sasakawa, C. and Yoshikawa, M. (1987) A series of Tn*5* variants with various drug-resistance markers and suicide vector for transposon mutagenesis. *Gene,* **56,** 283.
62. Joseph-Liauzum, E., Fellay, R., and Chandler, M. (1989) Transposable elements for efficient manipulation of a wide range of Gram-negative bacteria: promoter probes and vectors for foreign genes. *Gene,* **85,** 83.
63. Mahillon, J. and Kleckner, N. (1992) New IS*10* transposition vectors based on a Gram-positive replication origin. *Gene,* **116,** 69.
64. Metzger, M., Bellemann, P., Schwartz, T., and Geider, K. (1992) Site-directed and

transposon-mediated mutagenesis with pfd-plasmids by electroporation of *Erwinia amylovora* and *Escherichia coli* cells. *Nucleic Acids Res.*, **20**, 2265.
65. Haas, R., Kahrs A. F., Facius, D., Allmeier, H., Schmitt, R., and Meyer, T. F. (1993) Tn*Max*—a versatile mini-transposon for the analysis of cloned genes and shuttle mutagenesis. *Gene*, **130**, 23.
66. Berg, D. E., Davies, J., Allet, B., and Rochaix, J.-D. (1975) Transposition of R factor genes to bacteriophage lambda. *Proc. Natl Acad. Sci. USA*, **72**, 3628.
67. Phadnis, S. H., Huang, H. V., and Berg, D. E. (1989) Tn*5supF*, a 264-base-pair transposon derived from Tn*5* for insertion mutagenesis and sequencing DNAs cloned in phage lambda. *Proc. Natl Acad. Sci. USA*, **86**, 5908.
68. Kurnit, D. M. and Seed, B. (1990) Improved genetic selection for screening bacteriophage libraries by homologous recombination *in vivo*. *Proc. Natl Acad. Sci. USA*, **87**, 3166.
69. Simon, R., Priefer, U., and Puhler, A. (1983) A broad host range mobilization system for *in vivo* genetic engineering: transposon mutagenesis in Gram negative bacteria. *Biol. Technology*, **1**, 784.
70. Kasai, H., Isono, S., Kitakawa, M., Mineno, J., Akiyama, H., Kurnit, D. M., Berg, D. E., and Isono, K. (1992) Efficient large-scale sequencing of the *Escherichia coli* genome: implementation of a transposon- and PCR-based strategy for the analysis of ordered lambda phage clones. *Nucleic Acids Res.*, **20**, 6509.
71. Berg, D. E., Schmandt, M., and Lowe, J. B. (1983) Specificity of transposon Tn*5* insertion. *Genetics*, **105**, 813.
72. Moyed, H. S. and Bertrand, K. P. (1983) Mutations in multicopy Tn*10 tet* plasmids that confer resistance to inhibitory effects of inducers of *tet* gene expression. *J. Bacteriol.*, **155**, 557.
73. Berg, C. M., Liu, L., Wang, B., and Wang, M.-D. (1988) Rapid identification of bacterial genes that are lethal when cloned on multicopy plasmids. *J. Bacteriol.*, **170**, 468.
74. Kleckner, N., Roth, J., and Botstein, D. (1977) Genetic engineering *in vivo* using translocatable drug-resistance elements: new methods in bacterial genetics. *J. Mol. Biol.*, **116**, 125.
75. Seifert, H. S., Chen, E. Y., So, M., and Heffron, F. (1986) Shuttle mutagenesis: a method of transposon mutagenesis for *Saccharomyces cerevisiae*. *Proc. Natl Acad. Sci USA*, **83**, 735.
76. Davies, C. J. and Hutchison, C. A. (1991). A directed DNA sequencing strategy based upon Tn*3* transposon mutagenesis: application to the ADE1 locus on *Saccharomyces cerevisiae* chromosome I. *Nucleic Acids Res.*, **19**, 5731.
77. Grinsted, J., Cruz, F. D. L., and Schmitt, R. (1990) The Tn*21* subgroup of bacterial transposable elements. *Plasmid*, **24**, 163.
78. Berg, C. M., Liu, L., Coon, M., Strausbaugh, L. D., Gray, P., Vartak, N. B., Brown, M., Talbot, D., and Berg, D. E. (1989) pBR322-derived multicopy plasmids harboring large inserts are often dimers in *Escherichia coli* K-12. *Plasmid*, **21**, 138.
79. Taylor, A. L. (1963) Bacteriophage-induced mutation in *E. coli*. *Proc. Natl Acad. Sci. USA*, **50**, 1043.
80. van Gijsegem, F., Toussaint, A., and Casadaban, M. (1987) Mu as a genetic tool. In *Phage Mu*. N. Symonds, A. Toussaint, P. van de Putte, and M. M. Howe, (eds). Cold Spring Harbor Laboratory Press; Cold Spring Harbor, NY, p. 215.
81. Wang, B., Liu, L., Groisman, E. A., Casadaban, M. J., and Berg, C. M. (1987) High frequency generalized transduction by miniMu plasmid phage. *Genetics*, **116**, 201.

82. Wang, M.-D., Buckley, L., and Berg, C. M. (1987) Cloning of genes that suppress an *Escherichia coli* K-12 alanine auxotroph when present in multicopy plasmids. *J. Bacteriol.*, **169,** 5610.
83. Darzins, A. and Casadaban, M. J. (1989) Mini-D3112 bacteriophage transposable elements for genetic analysis of *Pseudomonas aeruginosa. J. Bacteriol.*, **171,** 3909.
84. Ahmed, A. (1984) Use of transposon-promoted deletions in DNA sequence analysis. *J. Mol. Biol.*, **178,** 941.
85. Wang, G., Berg, C. M., Chen, J., Young, A. C., Blakesley, R. W., Lee, L. Y., and Berg, D. E. (1993) Creating nested deletions for sequencing cosmid DNAs. *Focus,* **15,** 47.
86. Ahmed, A. (1987) A simple and rapid procedure for sequencing long (40-kb) DNA fragments. *Gene,* **61,** 363.
87. Slauch, J. M. and Silhavy, T. J. (1991) Genetic fusions as experimental tools. *Methods Enzymol.*, **204,** 213.
88. Manoil, C. and Beckwith, J. (1985) Tn*phoA*: a transposon probe for protein export signals. *Proc. Natl Acad. Sci. USA,* **82,** 8129.
89. Mielke, D. L. and Russel, M. (1992) A modified Tn*phoA* useful for single-stranded DNA sequencing. *Gene,* **118,** 93.
90. Albano, M. A., Arroyo, J., Eisenstein, B. I., and Engleberg, N. C. (1992) *PhoA* gene fusions in *Legionella pneumophila* generated *in vivo* using a new transposon, Mud*phoA*. *Mol. Microbiol.*, **6,** 1829.
91. Chow, W.-Y. and Berg, D. E. (1988) Tn5tac1, a derivative of transposon Tn*5* that generates conditional mutations. *Proc. Natl Acad. Sci. USA,* **85,** 6468.
92. Cookson, B. T., Berg, D. E., and Goldman, W. E. (1990) Mutagenesis of *Bordetella pertussis* with transposon Tn*5tac1*: conditional expression of virulence-associated genes. *J. Bacteriol.*, **172,** 1681.
93. Neuwald, A. F., Krishnan, B. R., Brikun, I., Kulakauskas, S., Suziedelis, K., Tomcsanyi, T., Leyh, T. S., and Berg, D. E. (1992) *cysQ,* a gene needed for cysteine synthesis in *Escherichia coli* K-12 only during aerobic growth. *J. Bacteriol.*, **174,** 415.
94. Neuwald, A. F., Krishnan, B. R., Ahrweiler, P. M., Frieden, C., and Berg, D. E. (1993) Conditional dihydrofolate reductase deficiency due to transposon Tn*5ac1* insertion downstream from the *folA* gene in *Escherichia coli. Gene,* **125,** 69.
95. de Lorenzo, V., Eltis, L., Kessler, B., and Timmis, K. N. (1993) Analysis of *Pseudomonas* gene products using *lacI*^{q}/P*trp-lac* plasmids and transposons that confer conditional phenotypes. *Gene,* **123,** 17.
96. Herrero, M., deLorenzo, V., and Timmis, K. N. (1990) Transposon vectors containing non-antibiotic resistance selection markers for cloning and stable chromosomal insertion of foreign genes in gram-negative bacteria. *J. Bacteriol.*, **172,** 6557.
97. McClelland, M., Jones, R., Patel, Y., and Nelson, M. (1987) Restriction endonucleases for pulsed field mapping of bacterial genomes. *Nucleic Acids Res.*, **15,** 5985.
98. Wong, K. K. and McClelland, M. (1992) Dissection of the *Salmonella typhimurium* genome by use of a Tn*5* derivative carrying rare restriction sites. *J. Bacteriol.*, **174,** 3807.
99. Kohara, Y., Akiyama, K., and Isono, K. (1987). The physical map of the whole *E. coli* chromosome: application of a new strategy for rapid analysis and sorting of a large genomic library. *Cell,* **50,** 495.
100. Studier, F. W. (1989) A strategy for high-volume sequencing of cosmid DNAs: random and directed priming with a library of oligonucleotides. *Proc. Natl Acad. Sci. USA,* **86,** 6917.

101. Rudd, K. E. (1993) Maps, genes, sequences, and computers: an *Escherichia coli* case study. *ASM News,* **59,** 335.
102. Seifert, H. S., Ajioka, R. S., Paruchuri, D., Heffron, F., and So, M. (1990) Shuttle mutagenesis of *Neisseria gonorrhoeae*: pilin null mutations lower DNA transformation competence. *J. Bacteriol.,* **172,** 40.
103. Boyle-Vavra, S. and Seifert, H. S. (1993) Shuttle mutagenesis: two mini-transposons for gene mapping and for *lacZ* transcriptional fusions in *Neisseria gonorrhoeae. Gene,* **129,** 51.
104. Labigne, A., Courcoux, P., and Tompkins, L. (1992) Cloning of *Campylobacter jejuni* genes required for leucine biosynthesis, and construction of *leu*-negative mutant of *C. jejuni* by shuttle transposon mutagenesis. *Res. Microbiol.,* **143,** 15.
105. Ubben, D. and Schmitt, R. (1986) Tn*1721* derivatives for transposon mutagenesis, restriction mapping and nucleotide sequence analysis. *Gene,* **41,** 145.
106. Ubben, D. and Schmitt, R. (1987) A transposable promoter and transposable promoter probes derived from Tn*1721*. *Gene,* **53,** 127.
107. Berg, D. E. and Berg, C. M. (1983) The prokaryotic transposable element Tn*5*. *Bio/ Technology,* **1,** 417.
108. Nag, D. K., Huang, H. V., and Berg, D. E. (1988) Bidirectional chain termination nucleotide sequencing: transposon Tn*5seq1* as a mobile source of primer sites. *Gene,* **64,** 135.
109. Kroos, L. and Kaiser, D. (1984) Construction of Tn*5lac*, a transposon that fuses *lacZ* expression to exogenous promoters, and its introduction into *Myxococcus xanthus. Proc. Natl Acad. Sci. USA,* **81,** 5816.
110. Manoil, C. (1990) Analysis of protein localization by use of gene fusions with complementary properties. *J. Bacteriol.,* **172,** 1035.
111. de Lorenzo, V., Herrero, M., Jakubzik, U., and Timmis, K. N. (1990) Mini-Tn*5* transposon derivatives for insertion mutagenesis promoter probing, and chromosomal insertion of cloned DNA in gram-negative eubacteria. *J. Bacteriol.,* **172,** 6568.
112. Guzzo, A. and DuBow, M. S. (1991) Construction of stable, single-copy luciferase gene fusions in *Escherichia coli. Arch. Microbiol.,* **156,** 444.
113. Wolk, C. P., Cai, Y., and Panoff, J. (1991) Use of a transposon with luciferase as a reporter to identify environmentally responsive genes in a cyanobacterium. *Proc. Natl Acad. Sci. USA,* **88,** 5355.
114. Phadnis, S. H., Kulakauskas, S., Krishnan, B. R., Hiemstra, J., and Berg, D. E. (1991) Transposon Tn*5supF*-based reverse genetic method for mutational analysis of *Escherichia coli* with DNAs cloned in lambda phage. *J. Bacteriol.,* **173,** 896.
115. Suwanto, A. and Kaplan, S. (1992) Chromosome transfer in *Rhodobacter sphaeroides*: Hfr formation and genetic evidence for two unique circular chromosomes. *J. Bacteriol.,* **174,** 1135.
116. Stojiljkovi, I., Trgovcevic, Z., and Salaj-Smic, E. (1991) Tn*5-rpsL*: a new derivative of transposon Tn*5* useful in plasmid curing. *Gene,* **99,** 101.
117. Ernst, A., Black, T., Yuping, C., Panoff, J.-M., Tiwari, D. N., and Wolk, C. P. (1992) Synthesis of nitrogenase in mutants of the cyanobacterium *Anabaena* sp. strain PCC 7120 affected in heterocyst development of metabolism. *J. Bacteriol.,* **174,** 6025.
118. Woodworth, D. L. and Kreuzer, K. N. (1992) A system of transposon mutagenesis for bacteriophage T4. *Molec. Microbiol.,* **6,** 1289.
119. Davis, R. W., Botstein, D., and Roth, J. R. (1980) *A Manual for Genetic Engineering: Advanced Bacterial Genetics.* Cold Spring Harbor Laboratory Press, Cold Spring Harbor, NY.

120. Way, J. C., Davis, M. A., Morisato, D., Roberts, D. E., and Kleckner, N. (1984) New Tn*10* derivatives for transposon mutagenesis and for construction of *lacZ* operon fusions by transposition. *Gene*, **32,** 369.
121. Symonds, N., Toussaint, A., van de Putte, P., and Howe, M. M. (eds) (1987) *Phage Mu*. Cold Spring Harbor Laboratory Press, Cold Spring Harbor, NY.
122. Youderian, P., Sugiono, P., Brewer, K. L., Higgins, N. P., and Elliott, T. (1988) Packaging specific segments of the *Salmonella* chromosome with locked-in Mud-P22 prophages. *Genetics*, **118,** 581.
123. Bremer, E., Silhavy, T. J., Weisemann, J. M., and Weinstock, G. M. (1984) Lambda p*lac*Mu: a transposable derivative of bacteriophage lambda for creating *lacZ* protein fusions in a single step. *J. Bacteriol.*, **158,** 1084.
124. Shapiro, J. A. (1979) Molecular model for the transposition and replication of bacteriophage Mu and other transposable elements. *Proc. Natl Acad. Sci. USA*, **76,** 1933.
125. Krishnan, B. R., Kersulyte, D., Brikun, I., Berg, C. M., and Berg, D. E. (1991) Direct and crossover PCR amplification to facilitate Tn*5supF*-based sequencing of lambda phage clones. *Nucleic Acids Res.*, **19,** 6177.

4 Eukaryotic transposable elements as tools to study gene structure and function

KIM KAISER, JOHN W. SENTRY, and DAVID J. FINNEGAN

1. Introduction

Eukaryotic genomes were once considered invariant. The discovery that they contain transposable elements, powerful agents of genetic change, showed this not to be the case. Transposition into coding sequence is an obvious source of variability. Transposable elements also cause chromosome rearrangement, due to a variety of mechanisms including imprecise excision and exchange between elements at different locations. Although such events normally occur in a more or less random manner, and at low frequencies, it is possible to harness transposable elements as tools for genome manipulation. This is best illustrated by the *P* element of *Drosophila*. In our discussion of *P*, and of similar elements from plants, we hope to indicate what might be achieved in any eukaryotic organism of interest.

1.1 Eukaryotic transposable elements

Transposable elements can represent a substantial fraction of a eukaryotic genome. They occur in families, members of which are generally dispersed. The size of a family can vary from fewer than ten to several hundred thousand, depending on the element and species concerned. On the basis of structure and presumed mechanism of transposition (Fig. 1), transposable elements fall into two broad classes: elements that transpose by reverse transcription of an RNA intermediate (Class I elements) and elements that transpose directly from DNA to DNA (Class II elements).

There are two types of Class I element. Class I.1 elements resemble retroviruses in having long terminal direct repeats (LTRs) and the potential to encode polypeptides similar to retroviral *gag* and *pol* products. Here we refer to them as retrovirus-like elements. Class I.2 elements have similar coding potential, but lack terminal repeats. They have A-rich domains at the 3′ end of the 'coding' strand. We refer to elements of this type as LINE-like elements after the first examples, the mammalian LINE or L1 elements.

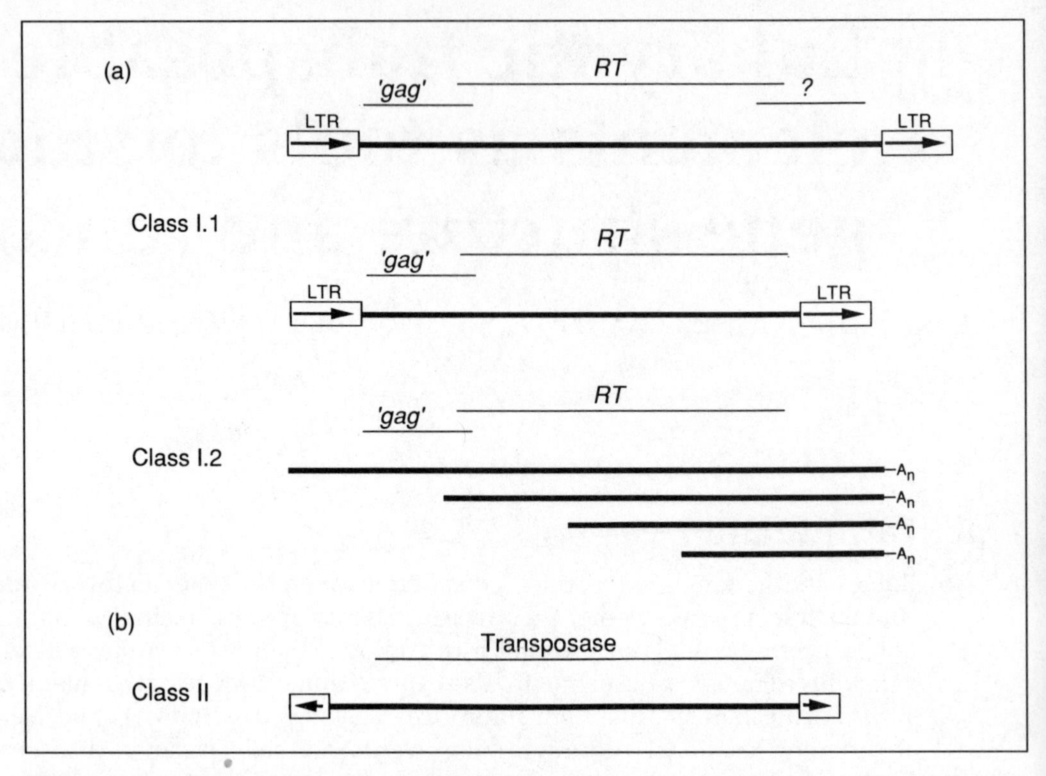

Fig. 1 Schematic representation of the structure of eukaryotic transposable elements. (a) Class I elements are of two types, those that are retrovirus-like, Class I.1, and those that are not, Class I.2. All Class I.1 elements have a gag-like ('*gag*') ORF, and an ORF that encodes a reverse transcriptase (*RT*). Some also have a third ORF of unknown function. The long terminal repeats (LTRs) are also shown. Class I.2 elements only have '*gag*' and *RT* ORFs. These elements have A-rich sequences (A_n) at their 3′ ends. Within a family many copies truncated at their 5′ ends are commonly found. (b) Class II elements have short terminal inverted repeats (arrowed). They contain a gene (which may have more than one exon) encoding transposase, an enzyme required for their own transposition.

Class II elements generally have inverted terminal repeats 10–200 bp long, although those of the *Drosophila FB* family are much longer. Class II elements encode proteins known as transposases. These have a direct role in the transposition process.

1.2 Mechanisms of transposition

1.2.1 Class I elements transpose via an RNA intermediate

Retrovirus-like elements are believed to transpose by a mechanism similar to that of retroviruses themselves. This has been established directly for the Ty*1* element of *Saccharomyces cerevisiae* (1). RNA transcripts initiating in one LTR and terminating in the other are assembled into virus-like particles together with proteins encoded by *gag*- and *pol*-like open reading frames (ORFs) (2). Element-encoded proteins generate a linear double-stranded DNA copy of the RNA by reverse transcription, and also play a role in integration at new sites. That RNA is indeed the transposition intermediate has been demonstrated by the precise removal during Ty*1* transposition of an intron that had been introduced into it. A similar observation has been made for *copia*, a retrovirus-like element of *Drosophila* (3).

Although LINE-like elements lack LTRs, they are also believed to transpose by a mechanism involving reverse transcription. Evidence for an RNA intermediate has been provided for the *I* element of *Drosophila* (4, 5), and for mouse L1 (6), again by showing that introns introduced into the elements can be removed precisely during transposition. By analogy with retrovirus-like elements, the reverse transcriptase of LINE-like elements is likely to be encoded by their *pol*-like ORFs. In the case of *jockey*, a LINE-like element of *Drosophila*, reverse transcriptase activity has indeed been demonstrated following expression in *Escherichia coli* (7). A less direct method has been used to demonstrate reverse transcriptase activity associated with the *CRE* element of *Trypanosoma brucei* (8), with the human *L1* element (9), and with the *Drosophila I* element (T. Paterson, A. Gabriel, J. Boeke, and D. J. Finnegan, unpublished).

The RNA transposition intermediate of a LINE-like element must include its entire sequence. Otherwise the element would be gradually truncated during successive rounds of transposition. One might expect the promoters from which transposition intermediates are synthesized to be internal to the elements. This is the case for the *Drosophila* elements *jockey* (10), *F* (11), and *I* (12).

1.2.2 Class II elements transposase without an RNA intermediate

A great deal has been learnt recently concerning the mechanism by which Class II elements transpose. It is primarily a conservative excision–insertion process. This is discussed in detail in Chapter 2, and is touched on in our discussion of the *Drosophila P* element (Section 2.1).

2. The *Drosophila P* element

2.1 Basic biology of the *P* element

Of all eukaryotic transposable elements, the most heavily exploited has been the *Drosophila P* element, a Class II element. Biological effects of *P* element transposition attracted the attention of researchers well before the causative agent was identified by molecular biologists. Crosses that we now know to mobilize *P* elements were observed to cause a diversity of effects including high rates of mutation, chromosomal rearrangement, male recombination (normally absent in *D. melanogaster*), and aberrant gonadal development with consequent sterility. The syndrome is known as hybrid dysgenesis (13). Its severity increases with temperature.

Hybrid dysgenesis occurs only as the result of a cross between males of a 'P strain' and females of an 'M strain'. The distinguishing characteristic of P strains is that they have full-length *P* elements. These encode their own transposase, and are thus autonomous. P strains can also contain defective, generally internally deleted, *P* elements. Transposition in a P strain is tightly regulated. It is repressed by a product of the full-length *P* element itself. The condition is known as 'P cytotype'. M strains, by contrast, lack autonomous *P* elements, and lay eggs that are permissive for *P* element transposition ('M cytotype'). The phenomenon of cytotype explains the restriction of hybrid dysgenesis to a cross between P males and M females. Only such a cross introduces active *P* elements into a background where they can transpose at significant rates. Since transposition is restricted to cells of the germline, phenotypic consequences are not observed until further generations.

The first *P* element to be cloned was a defective element, identified by virtue of having disrupted the *white* locus. The defective element was then used as a molecular probe to clone a complete element, confirmed as such on the basis of transpositional activity when injected into embryos of an M strain – it transposed from a plasmid into the *Drosophila* genome (14). Full-length elements have four exons encoding an 87 kDa transposase (Fig. 2). Because the third intron is not removed in somatic cells, transposase activity is restricted to the germline (15). An element with the third intron removed (Δ2,3) is able to transpose in somatic cells, and in addition lacks the capacity to establish P cytotype (16). *P* element sequences required in *cis* for transposition are contained within 138 bp at the 5′ end and 150 bp at the 3′ end (17). This includes a 31 bp terminal inverted repeat. Defective *P* elements, mainly internally deleted versions of the full-length element, can be found in both P and M strains. The extent of internal deletion varies. Defective elements usually retain *cis*-acting determinants that allow mobilization in the presence of full-length elements, even though they do not produce a functional transposase.

Transposition of a *P* element appears to involve its excision from the donor site. This leaves a double-strand break, repair of which is thought to be template-dependent (Fig. 3) (18, 19). Usually the template is provided by the sister chromatid. In such a case, *P* element sequences will seem to have been retained at the donor site. Internally deleted (defective) elements could result from incomplete repair. Loss of sequences flanking a *P* element, together with some or all of the element itself, would result from incomplete repair of a gap that had been widened by exonuclease activity. Rather than the sister chromatid, a homologous chromosome may on occasion be used as the template. Were it to carry a wild-type copy of the donor site (i.e. one lacking a *P* element), the impression would be given of precise excision.

2.2 Germline transformation

Introduction of cloned and manipulated genes into the germline DNA of an experimental organism is one of the most powerful tools of modern biology. Transpositional activity of a cloned *P* element following injection into M cytotype embryos paved the way for the generation of *P* element transformation vectors. Germline transformation of *Drosophila* is achieved by injecting plasmid DNA containing a

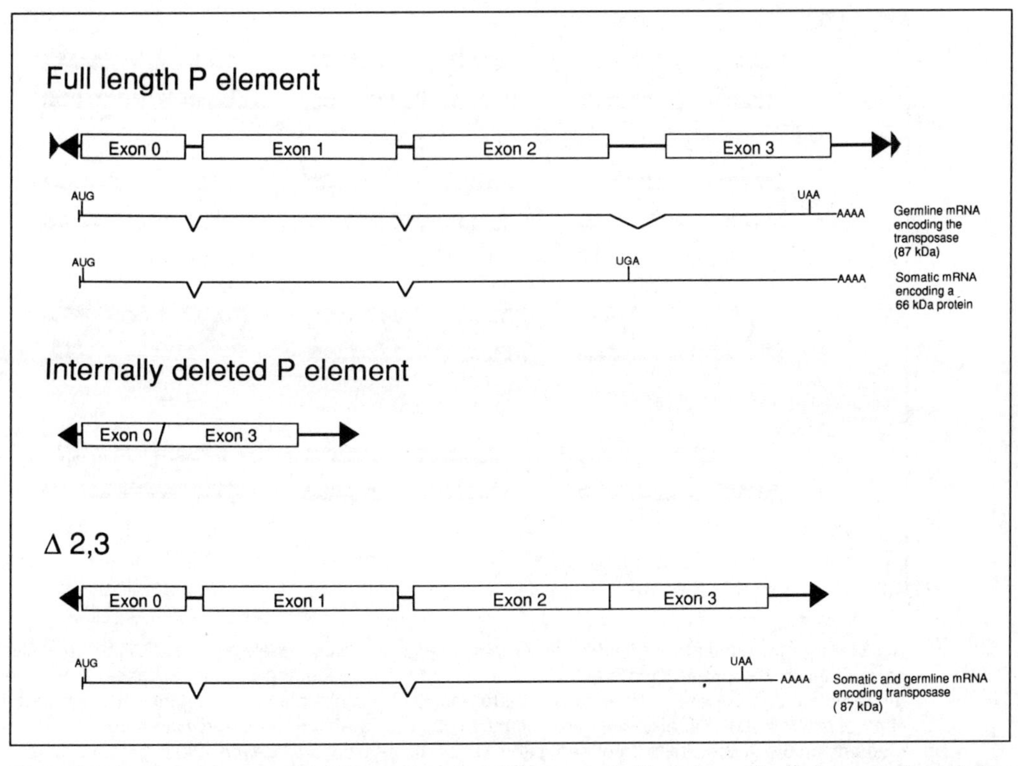

Fig. 2 Full-length *P* elements (2.9 kb) have four long ORFs separated by introns. The *P* element is bounded by 31 bp inverted repeats (large arrowheads). Insertion of a *P* element causes an 8 bp target site duplication (small arrowheads). Germline transcripts, spliced as shown, provide functional transposase. Somatic transcripts, which retain the intron between exons 2 and 3, encode a prematurely truncated and thus non-functional transposase. Only internally deleted *P* elements are present in the strain Birmingham 2. Such elements do not produce functional transposase and are thus non-autonomous, but they retain *cis*-acting determinants that allow their mobilization in the presence a transposase source. Δ2,3 elements, from which the third intron has been removed by *in vitro* manipulation, are often used to mobilize internally deleted *P* elements. Δ2,3 produces transposase in both germline and somatic tissues.

suitably manipulated *P* element into embryos undergoing the transition between syncitial and cellular blastoderm (14, 20). The earliest nuclei to cellularize are a group of germline progenitors that migrate to the posterior pole. *P* element DNA injected into that region can become internalized during cellularization, and can transpose to the genome. Transposition is not a frequent event on a *per* molecule basis, but none the less provides acceptable transformation efficiencies. Newly integrated elements will be inherited by the progeny of individuals that survive injection (Fig. 4). Transient expression has been observed following introduction of DNA into embryos by electroporation (21). Though not yet demonstrated, germline transformation may also be possible by such means.

An autonomous *P* element provides its own transposase. *P* elements engineered as vectors dispense with this ability, but retain sequences required in *cis* for transposition. In this respect they resemble the defective elements discussed

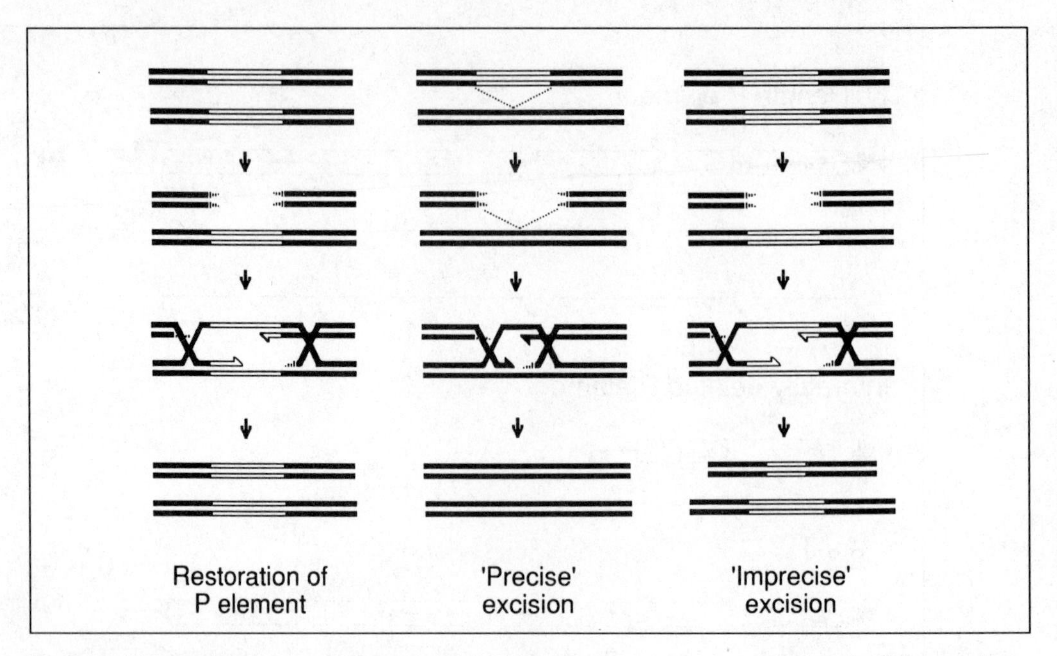

Fig. 3 Model for template-dependent gap repair following *P* element excision. Excision of a *P* element (open bars) induces a double-strand break that can be subject to widening by exonucleases. Free 3′ ends invade the template duplex, which serves as a substrate for DNA synthesis. In the left panel, the template is a second copy of the *P*-induced allele, most commonly provided by a sister chromatid. The result is restoration of a *P* element at the locus. Less frequently, the template can be a wild-type allele present on a homologous chromosome (centre panel). This will give the impression of precise excision. Interruption of the repair process, in this case where the sister chromatid is the template, followed by pairing of partially extended 3′ ends, may give the impression of an 'imprecise excision' (right panel). This can take the form of internal deletion of the *P* element or, more extremely, a deletion that extends into flanking DNA (most likely when the template is a wild-type allele present on a homologous chromosome).

above. It is therefore necessary to provide transposase from another source. Options are co-injection of an element that produces transposase but that cannot itself transpose—for example, a *wings-clipped* element (22), co-injection of purified transposase (23), or injection of the construct into embryos that express transposase endogenously, preferably from a Δ2,3 element which generates high levels of transposase activity without establishing P cytotype. In the latter case, generation of a line with a stable insertion of the construct requires selection against Δ2,3 in a subsequent generation. The frequency with which transformants are recovered appears inversely related to transposon length. None the less, transformation with cosmid-sized pieces of DNA can be achieved (24). There are even cosmid libraries constructed in *P* element vectors (25).

There can be pronounced position effects on the expression of genes contained within a *P* element transformation construct. It thus is advisable to generate a number of lines containing independent insertions. These can be generated either as primary transformants, or via remobilization of a construct by a cross that provides Δ2,3. It should be noted that *P* element transposition is non-random with

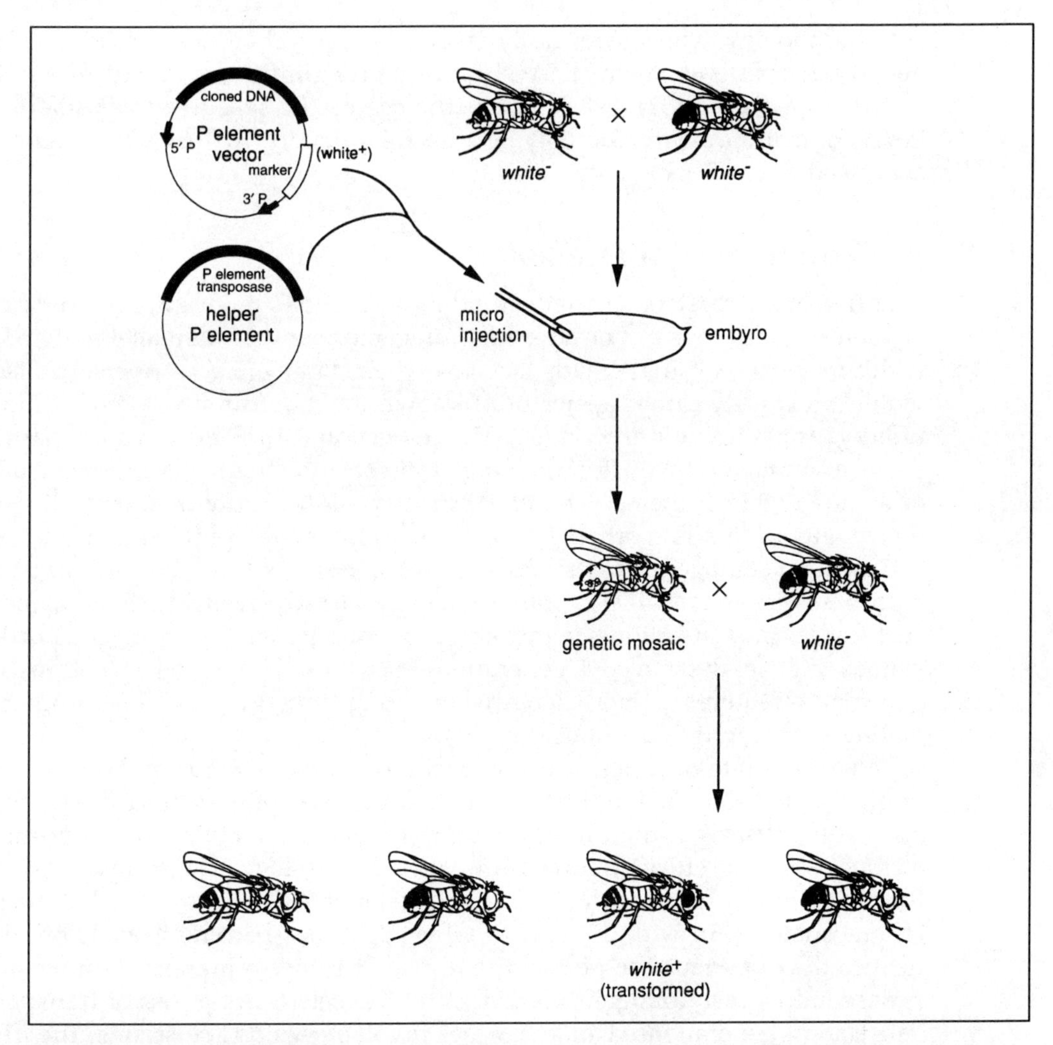

Fig. 4 Germline transformation. A *P* element vector, itself a component of a bacterial plasmid, contains the DNA one wishes to introduce into the *Drosophila* genome together with a marker gene that enables transformants to be recognized. Transposase can be supplied by co-injection of a plasmid that encodes it (*wings clipped* helper elements cannot themselves transpose), by co-injection of purified protein, or by injection of Δ2,3 embryos. DNA injected at the posterior pole prior to cellularization will become incorporated into germline precursors, and occasional transposition will occur from the injected plasmid to the *Drosophila* genome. Adults that develop from injected embryos are genetic mosaics with respect to the presence of transposon in their germline. It is therefore only in the next generation that 'fully-transformed' individuals can be recognized (here shown rescued by the marker for an eye-pigmentation defect).

respect to insertion site. Moreover, sequences contained within a *P* element construct can have a pronounced effect on insertion specificity (26).

A *P* element vector must carry a marker that enables successful transformation to be recognized. Markers that rescue a visible phenotypic defect, such as loss of eye colour (*rosy*, *white*, *vermilion*), loss of body pigmentation (*yellow*), or abnormal eye morphology (*rough*) are easily scored by eye (27–30). Alternatively, alcohol dehydrogenase and neomycin-resistance genes confer the ability to survive on selective media (31, 32). Markers can themselves be sensitive to position effects. Levels of marker expression may be a useful guide to whether a transgene will be expressed at reasonable levels.

2.3 *P* element mutagenesis

Any transposable element that can be mobilized at significant frequencies under laboratory conditions can be a useful mutagenic agent. Mutant alleles are provided with a molecular tag that greatly facilitates their cloning (33). Conversely, insertions within previously cloned genes of unknown function can be scored by molecular rather that phenotypic criteria (34, 35, see Section 2.4). *P* elements are particularly useful as mutagens because of their high transposition frequency and the availability of strains that lack them. The latter property allows backcrossing to eliminate all *P* elements from a line other than the one in the gene of interest.

P element mutagenesis is simplest in the context of the X chromosome, and of autosomal genes for which a mutant allele is already available (Fig. 5a and b). In such cases, new insertions can exhibit a phenotype while hemizygous or heterozygous (i.e. in the new F2 generation). Screening for novel mutations on the autosomes requires a more lengthy breeding strategy involving brother–sister mating within individual lines (Fig. 5c).

The overall rate of *P* element transposition clearly depends on both the number of mobilizable elements in the genome and the level of transposase activity. It has also been observed that transposition frequency is inversely proportional to element size. An efficient general mutagenesis strategy (Fig. 6) involves crossing Birmingham 2, a strain with 17 internally deleted *P* elements on each of its second chromosomes (36), with a strain in which Δ2,3 has become irreversibly inserted near to the dominant eye phenotype locus, *Dr* (37). An immobile source of transposase linked to a dominant marker simplifies selection for loss of transposase in subsequent generations. Unlike crosses involving wild-type strains, the direction of the above cross is irrelevant. Eggs laid by Δ2,3 females have M cytotype. One disadvantage of using Δ2,3 is transposase activity in the soma. This reduces the viability of dysgenic individuals. The problem can be minimized by performing the cross at a 16°C.

Each germ cell of a dysgenic individual is different with respect to its spectrum of new insertions (approximately 10 per haploid genome for strategies involving Δ2,3 and Birmingham 2) and it is necessary to go one further generation to obtain individuals from which a pure line can be bred. Were *P* element insertion completely random, 10 000 flies each with 10 new insertions would together contain a *P* element every 1 kb along the euchromatic genome (total size *c.* 10^5 kb). It cannot be guaranteed that any given target locus will have been the subject of insertion, however. The mutation rate depends on the gene in question. It can vary from $10^{-2}$to 10^{-7} of all progeny (13, 38).

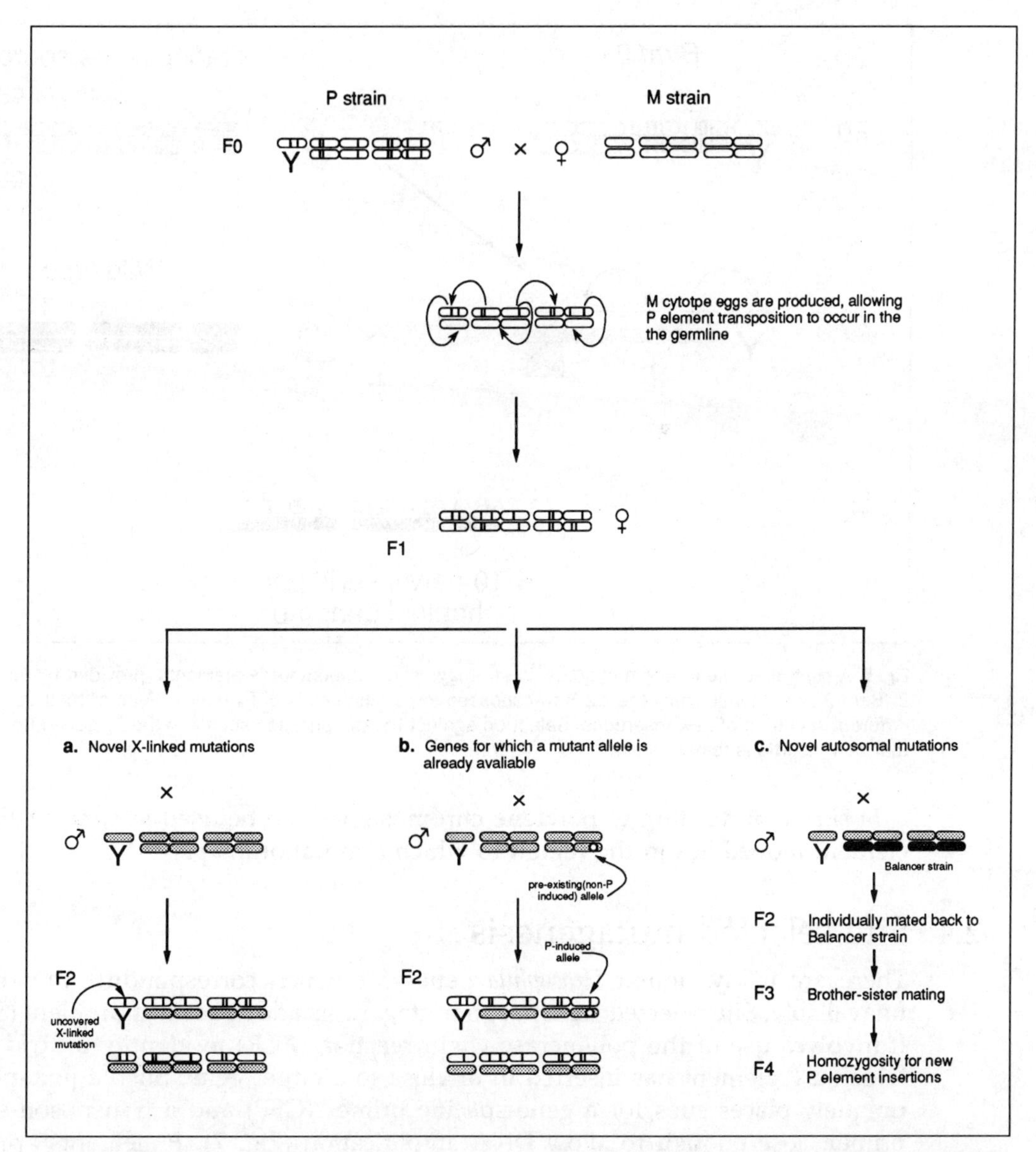

Fig. 5 *P* element mutagenesis. P strain males, carrying both autonomous and non-autonomous *P* elements, are mated with M strain females. The fertilized eggs are of M cytotype, allowing *P* element transposition to occur in the developing germline. As a result, each germline cell contains a new configuration of *P* elements. Phenotypic consequences are observed in subsequent generations. (a) Novel X-linked mutations are uncovered in F2 males. (b) Insertion within genes for which mutant alleles are already available are also recognizable in the F2. (c) Screening for *novel* mutations on the autosomes requires additional matings. New insertions must first be 'balanced' (a balancer chromosome carries a dominant visible marker, is lethal when homozygous, and suppresses homologous recombination: see ref. 27 for details). Brother–sister mating will eventually generate individuals homozygous for new *P* element insertions.

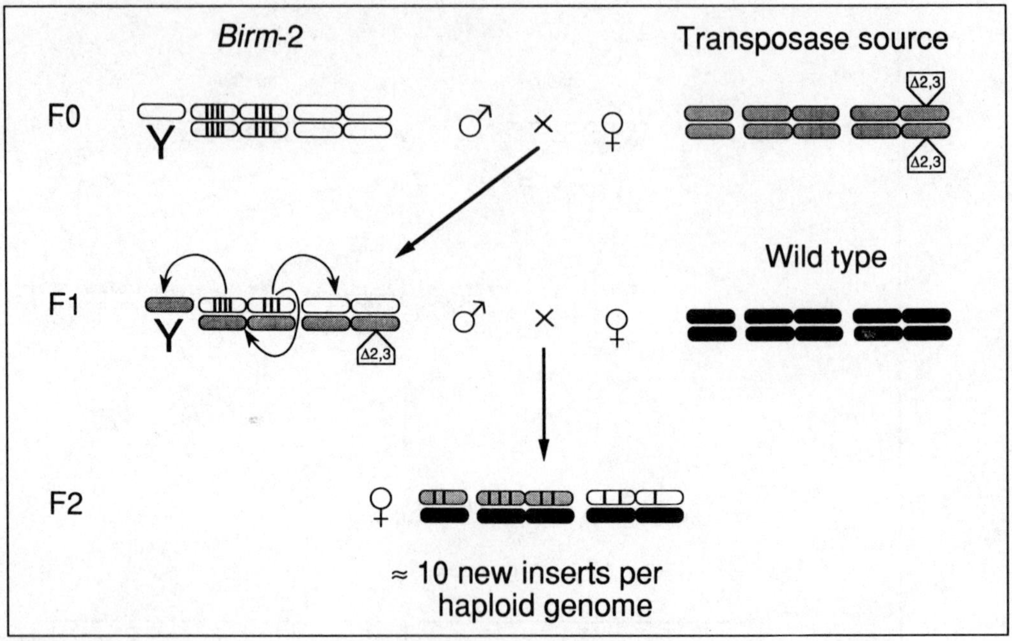

Fig. 6 A controlled *P* element mutagenesis strategy. Non-autonomous *P* elements, provided by Birmingham 2 (Birm-2), are mobilized by the Δ2,3 transposase in germline cells of F_1 males. Each of their sperm has a different spectrum of new insertions. Selection against the transposase source in the F_2 generation ensures that new insertions remain stable.

In situ hybridization to polytene chromosomes can be used to confirm that a *P* element indeed lies in the region to which a mutation maps.

2.4 Site-selected mutagenesis

There are many cloned *Drosophila* genes for which corresponding mutants are unavailable. Site-selected transposon mutagenesis addresses this problem (34, 35). It involves use of the polymerase chain reaction (PCR) to identify individuals in which a *P* element has inserted in or close to a target gene. Such a juxtaposition uniquely places sites for a gene-specific primer (GSP) and a transposon-specific primer close enough to allow DNA amplification (Fig. 7). *P* elements appear to insert preferentially at the 5′ ends of genes, so GSPs are most sensibly designed to target such regions. Since the sequence of a *P* element is bounded by 31 bp inverted repeats, a primer based on this sequence detects *P* elements inserted in either orientation with respect to the GSP.

The sensitivity of PCR allows rare individuals to be detected initially within a large population of mutagenized flies. They are then followed as specific amplification products while the population is sub-divided. Detection at the molecular rather than the phenotypic level facilitates fast and efficient screening and can be carried out on heterozygous individuals.

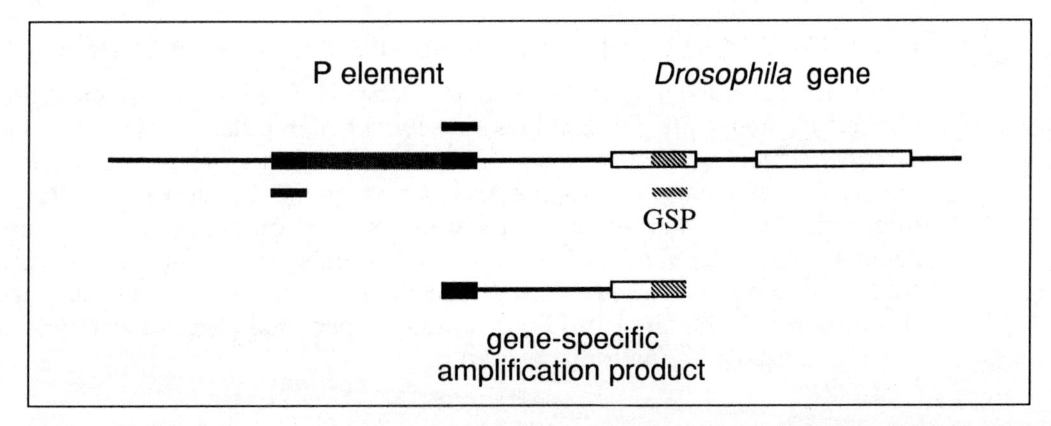

Fig. 7 Site-selected mutagenesis. Juxtaposition of a *P* element and a target gene uniquely provides a template for amplification between a gene-specific primer (GSP) and a transposon-specific primer based on the *P* element 31 bp inverted repeat. Open boxes represent exons of a hypothetical *Drosophila* gene.

As discussed above, the frequency of *P* element insertion can be highly gene-dependent. A summary of available data on site-selected mutagenesis is given in Table 1. Related strategies involve inverse or linker-mediated PCR for the amplification of sequences flanking all new *P* element insertions in a population of mutagenized individuals. Amplified DNA is then screened by hybridization with gene-specific probes (39).

Table 1 Site-selected *P* element mutagenesis of *Drosophila*. In all cases, Birmingham 2 was the source of mutagen, and Δ2,3 the source of transposase.

Target locus	Map position	Number of progeny scored	Number of insertions detected	Reference
singed	7D1-2	900	3	35
RI subunit of PKA	77F	12 000	2	–[a]
DCO (catalytic subunit of PKA)	30C1-6	7000	3	142
G_i–α (G protein)	65C	7000	3	142
G_o–α (G protein)	47A	7000	4	142
PKC	98F	7000	0	–[b]
Synaptobrevin	46E-F	7000	0	–[b]
Microtubule-associated protein 205K	100EF	15 000	5	143
72H5 (adult compound eye cDNA)		6316	1	34
4A11 (adult compound eye cDNA)		6316	1	34
pourquoi-pas?	98EF	40 000	3	144

[a] S. F. Goodwin, personal communication
[b] A. Duncanson, personal communication

2.5 Natural versus marked elements as general mutagens

There are several advantages to using marked *P* elements as mutagens (40). Marked elements greatly facilitate subsequent manipulation of a new allele, for example screening for excisions. Unfortunately marked elements are invariably much larger than natural elements, and in consequence transpose far less efficiently. In addition, most marked elements are present at only one or a few copies per genome, whereas strains are available that contain tens of natural elements. One strain has been constructed that has eight marked elements on an attached X chromosome (E. H. Grell and D. C. Lindsley, personal communication), but even this has a relatively low transposition rate.

2.6 Plasmid rescue

P elements engineered to contain a plasmid origin of replication and a drug-resistance determinant allow one-step recovery of *Drosophila* genomic DNA flanking the site of insertion. This procedure is known as plasmid rescue (Fig. 8) (41, 42). Such elements can also be used for a form of site-selected mutagenesis (43; E. Hafen, personal communication). A pool of plasmids rescued from a population of flies with different insertion sites contains sequences representing each flanking region. Hybridization between the pool and a specific cDNA/genomic DNA clone is diagnostic of an insertion in or near the gene of interest. The low frequency of transposition of marked elements tends to preclude their use as general mutagens. The technique is likely to prove useful in the context of local jumping, however (see below).

2.7 Local jumping

Recent evidence indicates that mobilization of *P* elements in the female germline leads to a high frequency of insertion within a hundred kilobases or so of the donor site (44, 45, 46). There is an ever increasing number of lines containing a single marked *P* element at a known chromosomal location. Hundreds are held by *Drosophila* stock centres and many more are being generated in the context of enhancer-trap screens and genome mapping programmes. Merely a thousand such lines place most of the genome within range of a local jump. *P* element transposition is not always accompanied by loss of the donor element (see Section 2.1). It may thus not be easy to score a local jump based on the marker that the transposon contains. Site-selected mutagenesis may be the most efficient approach.

2.8 Enhancer-trap elements

An enhancer-trap element is a modified *P* element containing a reporter gene with a very low level of intrinsic expression. This is due to the reporter being transcribed

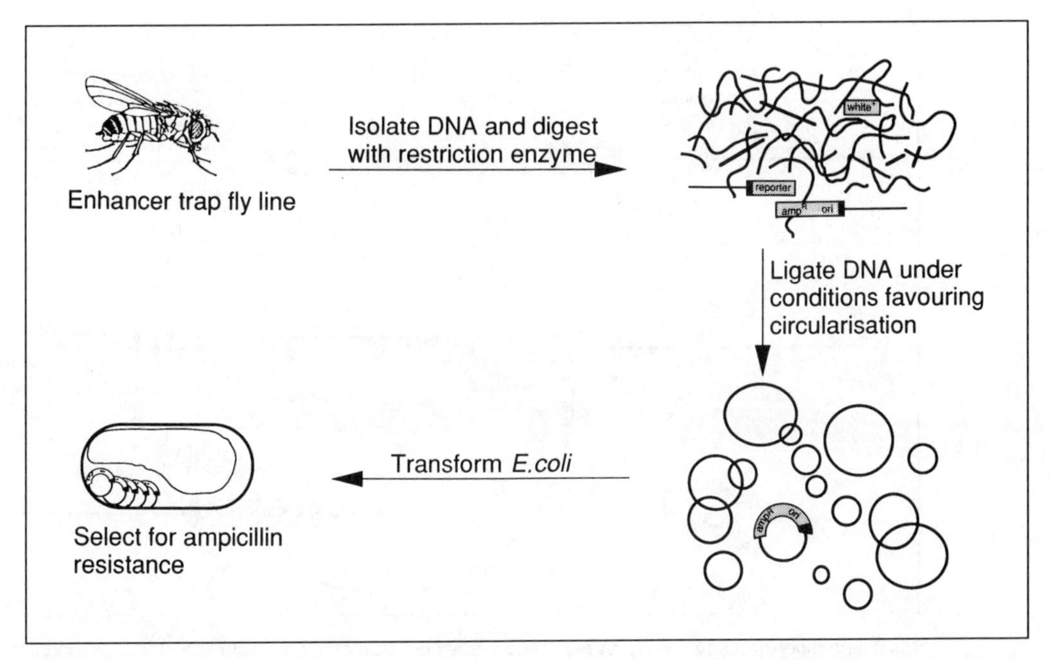

Fig. 8 Plasmid rescue. DNA is isolated from a line with a single engineered *P* element (here an enhancer-trap element) containing an *E. coli* origin of replication (*ori*) and a drug-resistance determinant (*amp*^{R}). The DNA is cleaved with an appropriate restriction enzyme, ligated under conditions that favour *intra*molecular ligation, and used to transform *E. coli*. Plasmids recovered from ampicillin-resistant colonies contain *Drosophila* DNA from adjacent to the site of *P* element insertion.

from a weak promoter lacking associated enhancer elements (47). Insertion into the *Drosophila* genome can lead to significant reporter expression, however, provided that it occurs in close proximity to a *Drosophila* enhancer. The pattern of expression of the reporter in a line with only one insertion thus reflects the temporal and spatial pattern of expression of a flanking *Drosophila* gene.

First generation enhancer-trap elements (Fig. 9a) contained the reporter gene *lacZ*, encoding the enzyme β-galactosidase. The presence of β-galactosidase activity in a tissue can be detected simply by its conversion of the chromogenic substrate X-gal. In addition to the reporter gene, enhancer-trap elements carry a marker gene such as *white* that enables flies with insertions to be recognized, and most include sequences that allow plasmid rescue of flanking DNA. One potential disadvantage of these elements is that they express β-galactosidase fused to the N-terminal nuclear localization signal of the *P* element transposase. Nuclear staining has its uses but precludes visualization of cell architecture, a particular problem in the study of cells with long processes, such as neurones.

A second generation enhancer-trap element has now been developed (Fig. 9b) (48) that can be used to target expression of any desired gene product to the marked cells. Instead of β-galactosidase it encodes the yeast transcription factor GAL4, which can function in *Drosophila* (49) to turn on expression of transgenes

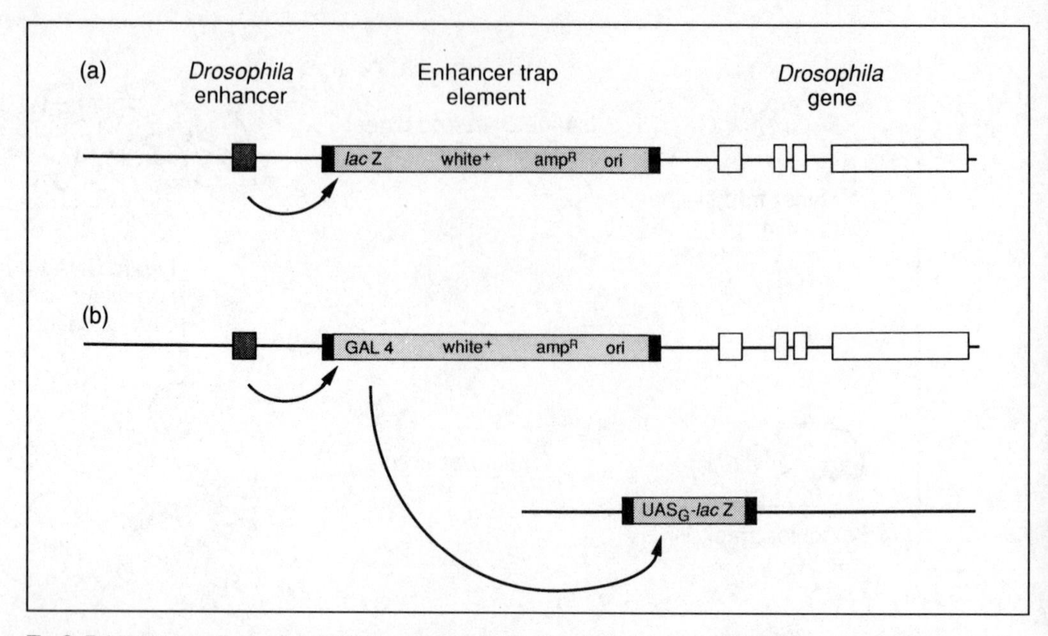

Fig. 9 Enhancer-trapping. (a) A first generation enhancer-trap element inserted within a *Drosophila* gene. The pattern and timing of expression of the reporter, *lacZ*, is dependent upon the specific genomic context in which it finds itself. *white*$^+$ is a marker that confers red eye colour in a *white*$^-$ genetic background, and thus allows flies containing new insertions to be recognized. The ampicillin resistance determinant (*amp*^{R}) and *E. coli* origin of replication (*ori*) facilitate plasmid rescue of flanking sequences. (b) A GAL4 enhancer trap element. The pattern and timing of GAL4 expression is similarly context dependent, and can be used to drive expression of a secondary reporter gene linked to the GAL4-responsive promoter, UAS_G.

placed under the control of a GAL4-dependent promoter (UAS_G). Crossing a fly having a new *GAL4* insertion with a fly containing a UAS_G–*lacZ* construct causes β-galactosidase to be expressed in a pattern that reflects GAL4 activity. In the case of β-galactosidase lacking a nuclear localization signal, this nicely side-steps the problem described above. An example of an enhancer-trap line showing region specific expression in the *Drosophila* brain is shown in Fig. 10.

Cell-type specific expression of a transcription factor provides a mean of expressing any gene product in the marked cells. Of considerable interest in this respect are cell ablation agents such as ricin and diphtheria toxin, and in particular temperature-sensitive versions (50, 51) that allow the timing of ablation to be controlled. Such strategies could be used to address a range of questions concerning the development and function of specific groups of cells (52, 53).

2.9 Precise and imprecise 'excision'

Reversion of a *P*-induced mutation by precise loss of the transposon may be the only unambiguous means of demonstrating that a phenotypic change is indeed the consequence of a lesion in a tagged or targeted gene. Such losses can be

Fig. 10 Cryostat section of a *Drosophila* brain showing GAL4-dependent expression of β-galactosidase in the mushroom bodies, regions thought to play a central role in learning and memory. Staining results from conversion of the chromogenic substrate, X-gal.

selected following remobilization of the *P* element, preferably from a background in which it is the only one remaining. Remobilization can also result in imprecise 'excision', leading to the generation of a range of new alleles of varying severity (54–56). Given that both site-selected mutagenesis and enhancer-trap experiments can generate lines in which *P* elements lie close, to rather than within, genes of interest, imprecise excision may be a necessary step in further analysis (57, 58). Excisions are much easier to score if marked elements are involved.

P element excision can also be used to generate chromosomal deficiencies. Deletions between two *P* elements on the same chromosome are recovered at high frequencies during a dysgenic cross (59). Strains containing closely linked pairs of *P* elements can be generated by local jumping. Since both *P* elements will carry the same marker, and one copy may remain, detection of such events may require a molecular strategy.

2.10 Directed genome modification

As discussed above, *P* element transposition leaves behind a double-strand break, repair of which appears to require a template (18). This suggests a means by which

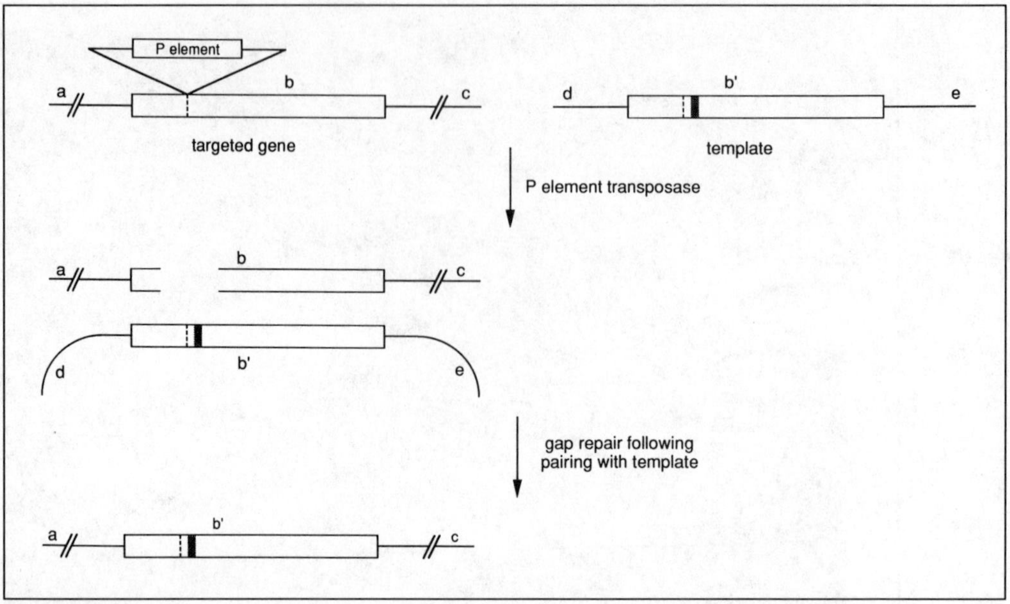

Fig. 11 Directed gene modification. The gene to be targeted (boxed region) contains a *P* element. In the presence of transposase the *P* element is excised, leaving behind a double-stranded gap that can be subject to widening by exonucleases (see Fig. 3). When the template for repair of the gap is an engineered copy of the gene (b′) the restored gene will contain the modification (shaded) present in the template.

specific modifications can be introduced into the *Drosophila* genome, i.e. by using a transgene as the template for repair (18, 19). The strategy works. Specific modifications made to a transgene can be introduced to a target locus (Fig. 11) (60, 61). These observations have considerable implications for any gene that has been the successful target of *P* element mutagenesis or of enhancer-trap insertion. Further discussion of this approach will be found in Chapter 2.

2.11 Can *P* elements be used in other species?

The extraordinary utility of *P* elements as tools with which to manipulate the *Drosophila* genome leads one to ask whether they might be used in similar ways in other species, particularly those that are of agricultural or medical importance, such as the Mediterranean fruity fly, *Ceritits capitata*, and mosquito vectors of malaria and other tropical and subtropical diseases. Successful integration of *P* element-based constructs in non-drosophiloid insects has yet to be reported (reviewed in ref. 62). A continuing line of *P* element research involves identifying host factors necessary for transposition, however (63). Such studies should shed light on the biological mechanisms that control dissemination and persistence of transposable elements, and may provide the key to the use of *P* in other species.

3. Other *Drosophila* transposable elements

Given the success of *P* elements as experimental tools in *Drosophila,* one might question whether there is any reason to develop transformation vectors, transposon tagging tools and enhancer-traps from other transposable elements. There are two reasons why this is justified; the non-random nature of *P* element insertion, and the apparent inability of *P* elements to integrate into the genomes of non-drosophiloid diptera and other species. Vectors developed from alternative transposable elements may be able to reach regions of the genome inaccessible to *P* elements, and may be less particular about where they are active.

3.1 Class II elements

3.1.1 *hobo*

Functional *hobo* elements have a structure similar to that of *P* elements. They are 3 kb long, have 12 bp terminal inverted repeats and contain a single 2 kb ORF that encodes a protein required for *hobo* transposition (64). *Drosophila* strains that contain 3 kb elements are called H strains. Those that do not are known as E strains. Flies produced by crossing H males and E females are genetically unstable. They have high frequencies of chromosome rearrangement, with *hobo* elements at the breakpoints (65–67). *hobo* elements can be used as transformation vectors in a similar way to *P* elements. A complete *hobo* element into which the *rosy* gene has been inserted is able to integrate into the genome after injection of embryos produced by crossing H males and E females, or after injection together with a complete element into embryos of an E strain (68).

The *hobo* transformation system has not been exploited to any great extent, but there may be circumstances in which it offers advantages. Transposon tagging and enhancer trapping with *hobo* may open up for manipulation regions of the genome that cannot be explored using *P*.

3.1.2 *mariner*

mariner is similar in structure to *P* and *hobo*. It was originally detected because of a somatically unstable insertion into the *white* gene of *D. mauritiana* (69). Somatic instability is a common feature of *mariner* transposition, unlike *P* and *hobo* for which genetic instability is largely confined to the germline. *mariner* is 1286 bp in length with 28 bp terminal inverted repeats and one long ORF coding for a peptide of 345 amino acids (70). Both autonomous elements capable of high rates of excision and transposition, such as *Mos1*, and non-autonomous elements that transpose in the presence of *Mos1*, have been identified. *mariner* has been used for transposon tagging in *D. mauritiana* (71).

mariner has been introduced into *D. melanogaster* by *P* element transformation, but can also transform this species unaided (72). It transposes more frequently in *D. melanogaster* than in *D. mauritiana*, and has a different site specificity from *P* and *hobo*, inserting preferentially at AT dinucleotides. It may well prove useful for targeting regions of the genome refractive to other transposons.

mariner-like elements have been found in a wide range of species. The first example from a non-drosophilid was cloned and sequenced from the cecropid moth *Hyalophora cecropia* (73). *mariner*-like elements have also been detected by a PCR-based strategy in 10 other species representing six additional orders (74). They are not restricted to insects, being found in a centipede, ascaid and phytoseiid mites (75), a fungus (75), the lower invertebrate planarian *Dugesia tigrina* (76), and in *Caenorhabditis elegans* (75, 77). The wide host distribution of *mariner* and its ability to transpose in another *Drosophila* species (78) suggests that it could be developed as a general agent for manipulating a wide range of animal species.

3.1.3 *minos*

Other Class II elements found in *Drosophila* may have potential for use as research tools. These include a family of transposable elements known as *minos* identified in *D. hydei*. *minos* is 1775 bp long with 255 bp long inverted terminal repeats and contains two non-overlapping ORFs with the potential to encode proteins of 153 and 201 amino acids (79). A marked *minos* element can transpose in *D. melanogaster* when injected into embryos together with a source of transposase. Its potential as a transformation vector for other Diptera is being investigated (B. Arca and C. Savakis, personal communication).

3.2 Class I elements

Class I transposable elements have not been exploited as experimental tools to the same extent as have Class II elements, although the *white* gene of *D. melanogaster* was cloned using an insertion of the retroviral-like element *copia* in the first transposon tagging experiment to be carried out in this species. If an element is to be useful for mutagenesis experiments of one sort or another then it must transpose at a respectable frequency and this is not generally the case. The *I* element of *D. melanogaster* is a LINE-like (Class I.2) element that transposes at high frequency during I–R hybrid dysgenesis (80). This occurs in the female progeny of crosses between males of a strain that contains a complete *I* element, an inducer strain, and females of a strain that does not, a reactive strain. *I* elements have been used successfully in a pilot site-selected mutagenesis experiment (81), but are not ideal for this purpose since all strains of *D. melanogaster* contain defective *I* elements, making further manipulation of a new allele problematic. Transposition of *I* elements is not associated with excision so phenotypic reversion cannot be used to confirm that a mutation is indeed associated with *I* element insertion, and imprecise excision cannot be used to produce secondary mutations. *I* elements are unlikely to be useful for germline transformation as transposition of defective elements is poorly complemented *in trans* (4, 5). The same may be true of other elements of this type.

3.3 A yeast recombinase system works in *Drosophila*

A plasmid-encoded recombination system from the yeast *S. cerevisiae* has been transferred to *Drosophila* with retention of function (82). It requires no components other than a single recombinase polypeptide (FLP) and its target sites (FRT sites), which can be as short as 34 bp (83). Where two FRT sites lie in close proximity on the same DNA molecule, FLP can cause inversion (sites in inverted orientation), or deletion (sites in direct orientation), of the intervening DNA. In the case of FRT sites on different DNA molecules, both reciprocal exchange and integrative recombination can result.

The FLP/FRT system has considerable potential. Examples from *Drosophila* include the following. Loss of a transcription termination site between two FRTs has been used as a way of turning genes on (Fig. 12a) (84). Somatic and germline mosaics have been generated by excision of a gene cloned between tandem FRT sites, FLP being provided conditionally under control of a heat shock promoter (Fig. 12b) (85–87). Generation of strains with FRT sites at the base of each chromosome arm has provided a general system for the production of somatic mosaics (Fig. 12c) (88). Rather than provide FLP under heat shock control, it could equally well be placed under GAL4 control in appropriate enhancer-trap lines.

Other recombinase systems requiring only a recombinase and specific target sites are available. In fact the bacteriophage P1-encoded Cre/lox recombination system (89), whose activity appears mechanistically identical to that of FLP (90, 91), is already being exploited in eukaryotes. To date, Cre/lox-mediated site-specific recombination has been reported in cultured mammalian cells (92), in transgenic mice (93, 94), and in transgenic plants (95, 96).

4. Transposable elements as tools in other organisms

4.1 *Caenorhabditis elegans*

Transposable elements have been described in several species of nematode (97), and sequences hybridizing with them have been detected across the phylum (98, 99). The elements isolated from *C. elegans* have been described in most detail, and of these Tc*1* is the best studied. It has been found in the genomes of all strains of *C. elegans* tested. The copy number and frequency of transposition varies widely between isolates (97). Tc*1* elements are 1610 bp long Class II elements with 54 bp terminal inverted repeats. The Tc*1* transposase is encoded by two exons and is 347 amino acids long. It stimulates transposition and excision in both the soma and germline (100). Transposition of Tc*1* appears to occur by excision–insertion, the double-strand break left after excision being healed by a gap repair mechanism similar to that hypothesized for *P* elements (101).

Tc*1* elements have been used for transposon tagging (102, 103) and site-selected mutagenesis (104). Frozen libraries of worms mutagenized by Tc*1* have been prepared for the latter purpose (105, 106). Tc*1* insertions in genes are most frequently

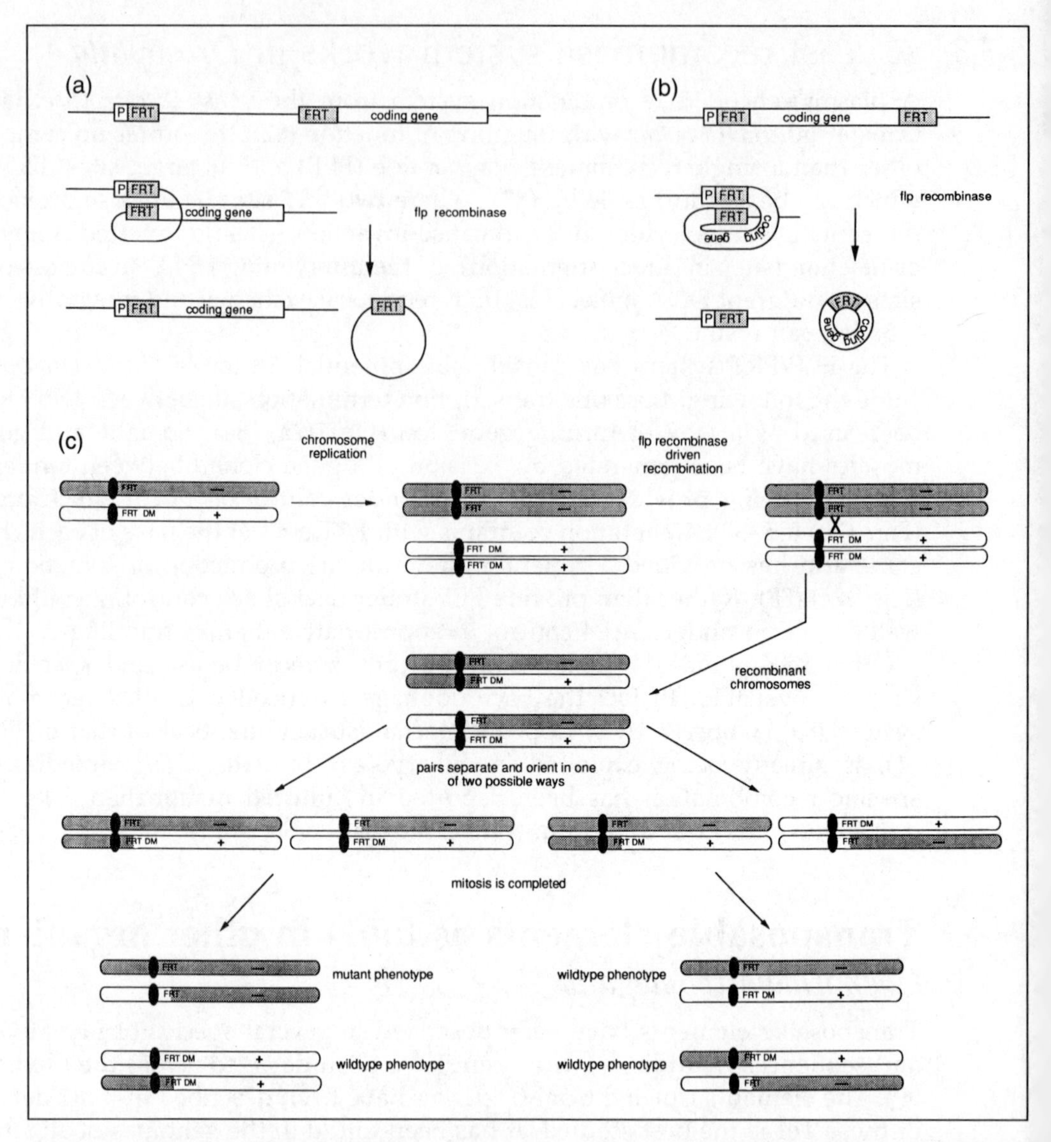

Fig. 12 Site-directed mitotic recombination in *Drosophila* using the yeast FLP/FRT system. (a) A transgene, separated from its promoter (P) as shown, is switched on by FLP-mediated deletion between the two FRT sites. FLP recombinase can be provided inducibly under the control of a heat-shock promoter. (b) A transgene flanked by FRT sites is 'silenced' by FLP-mediated deletion. (c) Placement of an FRT site at the base of a chromosome arm provides a general system for the production of somatic mosaics. One of a pair of such chromosomes carries a dominant cell-autonomous marker (DM; e.g. a transgene encoding a non-*Drosophila* epitope) distal to the FRT site. The other chromosome carries the mutation of interest (−). Heat-shock induction of FLP recombinase (from a construct on another chromosome) leads to recombination between FRT sites. Appropriate segregation of sister chromatids at the next cell division results in a cell that is homozygous for the mutation, and that lacks the dominant marker. Such a cell will found a patch of mutant tissue within an otherwise normal individual.

found in introns, resulting in little or no phenotypic consequence (107). Mobilization of Tc*1*, like that of *P* elements, can result in 'imprecise excision', generating deletion alleles for phenotypic study (108). The template-driven, double-strand gap repair, model described by Engels *et al.* (18) for *P* element excision is thought to be applicable to Tc*1* (101), and has been exploited for transgene-directed repair (109).

Tc3 is another element that may prove to be a useful tool in *C. elegans*. It is 2335 bp long with 462 bp terminal inverted repeats and two ORFs that are believed to be exons for the transposase gene (110). Most isolates of *C. elegans* have about 15 copies of Tc3 but so far there is no evidence for their transposition or excision (111). Transposition of the endogenous elements of the strain Bristol N2 can be induced by a Tc3 element under the control of a heat shock promoter, suggesting that Tc3 could be used for mutagenesis (110).

4.2 Plants

The first indication that eukaryotic genomes contain transposable genetic elements came from McClintock's studies of unstable mutations in maize (112). This led to the discovery of several families of Class II transposable elements in *Zea mays*. Within each family there are both autonomous elements that can stimulate their own transposition and non-autonomous elements that can only transpose in the presence of transposase from an autonomous element of the same family. Several systems of autonomous and non-autonomous elements are known but the best studied are the *Ac/Ds* family and *Spm/dSpm*, also known as *En/I* (113).

4.2.1 *Ac/Ds* (Activator/Dissociation) from *Zea mays*

Ac elements cause unstable mutations when they insert within or adjacent to genes because they can excise to give alleles that have wild-type or nearly wild-type activity. *Ds* elements can transpose and excise under the influence of *Ac* but are stable in strains that lack such elements. McClintock's genetic experiments suggested that *Ds* elements could be derived from *Ac* and this was confirmed when elements of each sort were recovered from unstable alleles of the *waxy* gene (114).

Ac elements are 4.6 kb long with 11 bp terminal inverted repeats and five exons that are joined by RNA splicing to give a 3.5 kb mRNA (115, 116) that can be translated into an 807 amino acid transposase. This is the only element-encoded product required for excision of *Ds* elements as excision of a *Ds* element in tobacco is stimulated by expression of transposase cDNA (117). Transposase binds to repeats of a hexameric sequence found in 200 bp regions that are adjacent to the 11 bp terminal inverted repeats (118) and that are essential for efficient transposition (119). The sequences of several *Ds* elements have been determined and in each case they appear to have been derived from *Ac* elements by internal deletions affecting one or more exons of the transposase gene (113). An exception is a 405 bp *Ds* element that is related to *Ac* only in its terminal inverted repeats (120).

Table 2 Summary of plants in which modified *Ds* elements, introduced using T-DNA transformation, were able to transpose and excise under the influence of *Ac* transposase

Plant	Subdivision	Family	Reference
Rice	Monocot	Poaceae	145
Wheat	Monocot	Poaceae	145
Nicotiana tabacum	Dicot	Solanaceae	146
Nicotiana plumbaginifolia	Dicot	Solanaceae	147
Tomato	Dicot	Solanaceae	148
Potato	Dicot	Solanaceae	149
Petunia	Dicot	Solanaceae	139
Carrot	Dicot	Umbelliferae	150
Parsley	Dicot	Umbelliferae	–[a]
Arabidopsis	Dicot	Brassicaceae	150
Soybean	Dicot	Fabaceae	151
Flax	Dicot	Linaceae	152

[a] R. Lutticke and R. Kunze, personal communication

Once *Ac* and *Ds* elements had been cloned they could be used as probes in transposon tagging experiments; several maize genes have been cloned in this way (121). This raised the possibility that they might be used in a similar way in other plant species into which these elements could be introduced using T-DNA transformation vectors from *Agrobacterium*. Modified *Ds* elements have been introduced into a wide range of dicotyledenous plants and in each case they have been shown to transpose and excise under the influence of *Ac* transposase (Table 2). This has encouraged several groups to screen for *Ds*-induced mutations affecting particular phenotypes. Initially this has to be done using donor elements introduced at random into the genome of the species of interest. These elements transpose preferentially to closely linked sites in maize and tobacco (122, 123) and if this is generally true the procedure could be refined to use strains with donor elements located at sites linked to the target.

A transposon tagging system like that in *Drosophila* requires both a source of transposase and transposing element. If these are separate then stable transposition events can be recovered by crossing out the source of transposase. Two component systems of this type have been developed using *Ac* transposase to drive transposition of *Ds* elements. The frequency with which these elements transpose varies from host to host and is particularly low in the model species *Arabidopsis thaliana*. This is due, in part at least, to the rate at which the transposase gene is transcribed and can be increased by linking the gene to the strong promoter of the cauliflower mosaic virus (CaMV) 35S gene (124). Mutations associated with *Ds* insertions have been isolated in *Arabidopsis* (125, 126) and *Petunia* (127) and no doubt other species will follow.

An enhancer-trap system for use in plants has also been developed from *Ac* and

Ds elements (128). The source of transposase is a modified *Ac* element with one of its inverted repeats deleted so that it cannot transpose itself and its transposase gene under the control of the CaMV 35S promoter. The enhancer trap element is a *Ds* element containing a promotorless β-glucuronidase (GUS) gene. This should give GUS activity in particular plant cells if it inserts downstream of a promoter and in the vicinity of appropriate enhancers. This can be detected using a chromogenic substrate similar to that used for β-galactosidase.

4.2.2 *Spm/dSpm* (suppressor–mutator) from *Zea mays*

The structure of *Spm/En* elements is similar to that of *Ac*. They are 8.3 kb long, have 13 bp terminal inverted repeats and encode mRNAs for two polypeptides, TNPA and TNPD, that are derived by alternative splicing from a precursor RNA (113, 129, 130). The TNPA transcript is the more abundant but both are required for transposition and excision (131, 132). The sequences required in *cis* for *Spm* transposition are located within 200 bp of the 5′ end and 300 bp of the 3′ end, regions that include repeats of a 12 bp motif that is the binding site for TNPA protein (133). These elements can also transpose in heterologous species (134) and have been used successfully for transposon tagging in maize and *Petunia* (121, 135).

4.2.3 Tam elements from *Antirrhinum majus*

Transposable elements in the snapdragon, *Antirrhinum majus* (Tam elements), have been studied for many years because they are the cause of variegated patterns of colour seen in the flowers of many strains; nine such elements have been identified so far. The best studied are Tam*1*, Tam*2*, and Tam*3* (136). Tam*1* and Tam*2* are related to each other and to maize *En/Spm* elements (137). They have related 13 bp terminal inverted repeats and Tam*1* and *En/Spm* have similar sequence organizations and encode similar polypeptides. Tam*3* is related to *Ac* in similar ways (138). It is an autonomous element that is particularly useful for mutagenesis and transposon tagging experiments as its transposition frequency is increased 1000-fold in plants propagated at 15°C as compared with the frequency at 25°C (136). Insertions of Tam elements have identified mutations in several genes some of which have been cloned using Tam sequences as probes (121). Tam*3* has been shown to transpose in tobacco and *Petunia* (139, 140) and may well prove to be as versatile as *Ac*.

5. Conclusions

The ideal transposable element system for use in eukaryotic genome analysis should have the following properties:

- a two-element system, separating the enzymatic function(s) from the target sequences of the transposable element
- a genetic background free of the transposable element

- a high rate of transposition
- little or no insertion site specificity
- an ability to mediate germline transformation
- an ability to generate precise and imprecise excision events
- an ability to use ectopic templates for gap repair following transposon excision

The properties of Class II elements have made it relatively easy to generate systems offering most of the above properties. Much of the technology for use of eukaryotic elements in gene analysis has been developed using the *Drosophila P* element. Its major present limitations are some insertion site specificity and an apparent inability of the transposase to work outside the drosophilids (62). Similar progress is being made in plants with *Ac, Spm,* and Tam*3*, and in *C. elegans* with Tc*1*. Class I elements have been less useful. The relative complexity of their transposition mechanisms will probably preclude their widespread use as experimental tools although a transformation vector has been developed from Ty*1* (141), and the same could be done for other retroviral elements provided that their transposition frequencies could be increased naturally or by experimental manipulation.

Transposable elements have great potential as tools with which to manipulate eukaryotic genomes, both for increasing our understanding of biological processes and for modifying organisms for practical purposes. Scientists working with species for which transposable element systems have not been developed have to choose between adapting an existing system or finding a suitable element to exploit within their own organism. Where possible the first route should be taken as it is likely that a considerable investment will be required before new elements can be usefully exploited, and an element taken from one species to another will usually be placed in a genome that has no related sequences. This simplifies subsequent manipulation and reduces the chance that transposition of the vector will be affected by an endogenous regulatory system.

Acknowledgements

We are grateful to Mingyao Yang for supplying the cryostat section of a *Drosophila* brain, and Reinhart Kunze for supplying the information in Table 2. Some of the results reported here are the result of work supported by the Human Frontier Science Program (D. J. F.), the Medical Research Council (D. J. F. and K. K.), and the Wellcome Trust (K. K.).

References

1. Boeke, J. D. (1989) Transposable elements in *Saccharomyces cerevisiae*. In *Mobile DNA*. D. E. Berg and M. M. Howe (eds). American Society for Microbiology, Washington, DC, p. 335.

2. Burns, N. R., Saibil, H. R., White, N. S., Pardon, J. F., Timmins, P. A., Richardson, S. M. H., Kingsman, S. M., and Kingsman, A. J. (1992) Symmetry, flexibility and permeability in the structure of yeast retrotransposon virus-like particles. *EMBO J.*, **11,** 1155.
3. Yoshioki, K., Kanda, H., Akiba, H., Enoki, M., and Shiba, T. (1991) Identification of an unusual structure in the *Drosophila melanogaster* transposable element *copia*: evidence for *copia* transposition through an RNA intermediate. *Gene,* **103,** 179.
4. Pélisson, A., Finnegan, D. J., and Bucheton, A. (1991) Evidence for retrotransposition of the I factor, a LINE element of *Drosophila melanogaster*. *Proc. Natl Acad. Sci. USA,* **88,** 4907.
5. Jensen, S. and Heidmann, T. (1991) An indicator gene for detection of germline retrotransposition in transgenic *Drosophila* demonstrates RNA-mediated transposition of the LINE *I* element. *EMBO J.*, **10,** 1927.
6. Evans, J. P. and Palmiter, R. D. (1991) Retrotransposition of the mouse L1 element. *Proc. Natl Acad. Sci. USA,* **88,** 8792.
7. Ivanov, V. A., Mellnikov, A., Siunov, A. V., Fodor, I. I., and Ilyin, Y. V. (1991) Authentic reverse transcriptase is coded by *jockey*, a mobile *Drosophila* element related to mammalian LINEs. *EMBO J.*, **10,** 2489.
8. Gabriel, A. and Boeke, J. D. (1991) Reverse transcriptase encoded by a retrotransposon from the trypanosomatid *Crythidia fasciculata*. *Proc. Natl Acad. Sci. USA,* **88,** 9794.
9. Mathias, S. L., Scott, A. F., Kazazian, H. H., Boeke, J. D., and Gabriel, A. (1991) Reverse transcriptase encoded by a human transposable element. *Science,* **254,** 1808.
10. Mizrokhi, L. J., Georgieva, S. G., and Ilyin, Y. V. (1988) *jockey*, a mobile *Drosophila* element similar to mammalian LINEs, is transcribed from the internal promoter by RNA polymerase II. *Cell,* **54,** 685.
11. Minchiotti, G. and DiNocera, P. P. (1991) Convergent transcription initiates from oppositely orientated promoters within 5′ end regions of *Drosophila melanogaster* F elements. *Mol. Cell. Biol.*, **11,** 5171.
12. McLean, C., Bucheton, A., and Finnegan, D. J. (1993) The 5′ untranslated region of the I factor, a long interspersed nuclear element-like retrotransposon of *Drosophila melanogaster*, contains an internal promoter and sequences that regulate expression. *Mol. Cell. Biol.*, **13,** 1042.
13. Engels, W. R. (1989) *P* elements in *Drosophila melanogaster*. In *Mobile DNA*. D. E. Berg and M. M. Howe (eds). American Society for Microbiology, Washington, DC, p. 437.
14. Spradling, A. C. and Rubin, G. M. (1982) Transposition of cloned *P* elements into *Drosophila* germ line chromosomes. *Science,* **218,** 341.
15. Rio, D. C. (1991) Regulation of *Drosophila P* element transposition. *Trends Genet.*, **7,** 282.
16. Laski, F. A., Rio, D. C., and Rubin, G. M. (1986) Tissue-specificity of *Drosophila P* element transposition is regulated at the level of mRNA splicing. *Cell,* **44,** 7.
17. Mullins, M. C., Rio, D. C., and Rubin, G. M. (1989) *Cis*-acting DNA sequence requirements for P element transposition. *Genes Dev.*, **3,** 729.
18. Engels, W. R., Johnson-Schlitz, D. M., Eggleston, W. B., and Sved, J. (1990) High-frequency P element loss in *Drosophila* is homolog dependent. *Cell,* **62,** 515.
19. Sentry, J. W. and Kaiser, K. (1992) P element transposition and targeted manipulation of the *Drosophila* genome. *Trends Genet.* **8,** 329.
20. Rubin, G. M. and Spradling, A. C. (1982) Genetic transformation of *Drosophila* with transposable element vectors. *Science,* **218,** 348.

21. Kamdar, P., von Allmen, G., and Finnerty, V. (1992) Transient expression of DNA in *Drosophila* via electroporation. *Nucleic Acids Res.*, **20,** 3526.
22. Karess, R. E. and Rubin, G. M. (1984) Analysis of *P* transposable element functions in *Drosophila*. *Cell*, **38,** 135.
23. Kaufman, P. D. and Rio, D. C. (1991) Germline transformation of *Drosophila melanogaster* by purified *P* element transposase. *Nucleic Acids Res.*, **19,** 6336.
24. Haenlin, M., Steller, H., Pirrotta, V., and Mohier, E. (1985) A 43 kilobase cosmid P transposon rescues the fs(1)K10 morphogenetic locus and three adjacent *Drosophila* developmental mutants. *Cell*, **40,** 827.
25. Speek, M., Raff, J. W., Harrison-Lavoie, K., Little, P. F. R., and Glover, D. M. (1988) Smart2, a cosmid vector with a phage lambda origin for both systematic chromosome walking and P element mediated gene transfer in *Drosophila*. *Gene*, **64,** 173.
26. Kassis, J. A., Noll, E., Vansickle, E. P., Odenwald, W. F., and Perrimon, J. (1992) Altering the insertional specificity of a *Drosophila* transposable element. *Proc. Natl Acad. Sci. USA*, **89,** 1919.
27. Ashburner, M. (1989) In *Drosophila: A Laboratory Handbook*. Cold Spring Harbor Laboratory Press, Cold Spring Harbor, NY, p. 1017.
28. Fridell, Y.-W. C. and Searles, L. L. (1991) *Vermilion* as a small selectable marker gene for *Drosophila* transformation. *Nucleic Acids Res.*, **19,** 5082.
29. Patton, J. S., Gomes, X. V., and Geyer, P. K. (1992) Position-independent germline transformation in *Drosophila* using a cuticle pigmentation gene as a selectable marker. *Nucleic Acids Res.*, **20,** 5859.
30. Lockett, T. J., Lewy, D., Holmes, P., Medveezky, K., and Saint, R. (1992) The *rough* (*ro*$^+$) gene is a dominant marker in germ line transformation of *Drosophila melanogaster*. *Gene*, **114,** 187.
31. Goldberg, D. A., Posakony, J. W., and Maniatis, T. (1983) Correct developmental expression of a cloned alcohol dehydrogenase gene transduced into the *Drosophila* germ line. *Cell*, **34,** 59.
32. Steller, H. and Pirrotta, V. (1985) A transposable P vector that confers selectable G418 resistance to *Drosophila* larvae. *EMBO J.*, **4,** 167.
33. Bingham, P. M. (1981) Cloning of DNA sequences from the *white* locus of *D. melanogaster* by a novel and general method. *Cell*, **25,** 693.
34. Ballinger, D. G. and Benzer, S. (1989) Targeted gene mutations in *Drosophila*. *Proc. Natl Acad. Sci. USA*, **86,** 4055.
35. Kaiser, K. and Goodwin, S. F. (1990) 'Site-selected' transposon mutagenesis of *Drosophila*. *Proc. Natl Acad. Sci. USA*, **87,** 1686.
36. Engels, W. R., Benz, W. K., Preston, C. R., Graham, P. L., Phillis, R. W., and Robertson, H. M. (1987) Somatic effects of P element activity in *Drosophila melanogaster*: pupal lethality. *Genetics*, **117,** 745.
37. Robertson, H. M., Preston, C. R., Phillis, R. W., Johnson-Schlitz, D. M., Benz, W. K., and Engels, W. R. (1988) A stable source of P element transposase in *Drosophila melanogaster*. *Genetics*, **118,** 461.
38. Kidwell, M. G. (1987) A survey of success rates using P element mutagenesis in *Drosophila melanogaster*. *Drosophila Inform. Serv.*, **66,** 81.
39. Sentry, J. W. and Kaiser, K. (1994) Application inverse PCR to site-directed mutagenesis of *Drosophila Nucl. Acids. Res.*, **16,** 3429.
40. Cooley, L., Berg, C., and Spradling, A. (1988) Controlling P element insertional mutagenesis. *Trends Genet.*, **4,** 254.

41. Pirrotta, V. (1986) Cloning *Drosophila* genes. In: *Drosophila: A Practical Approach*. D. B. Roberts (ed.). IRL Press, Oxford, p. 83.
42. Steller, H. and Pirrotta, V. (1986) P transposons controlled by heat shock promoter. *Mol. Cell. Biol.*, **6,** 1640.
43. Hamilton, B. A., Palazzolo, M. J., Chang, J. H., VijayRaghavan, K., Mayeda, C. A., Whitney, M. A., and Meyerowitz, E. M. (1991) Large scale screen for transposon insertions into cloned genes. *Proc. Natl Acad. Sci. USA*, **88,** 2731.
44. Tower, J., Karpen, G. H., Craig, N., and Spradling, A. C. (1993) Preferential transposition of *Drosophila P* elements to nearby chromosomal sites. *Genetics*, **133,** 347.
45. Zhang, P. and Spradling, A. C. (1993) Efficient and dispersed local P element transposition from *Drosophila* females. *Genetics*, **133,** 361.
46. Hamilton, B. A., Ho, A., and Zinn, K. (1994) Directed mutagenesis and genetic analysis of a *Drosophila* receptor-linked protein tyrosine phosphatase gene. *Roux's Arc. Dev. Biol.* (in press).
47. O'Kane, C. J. and Gehring, W. J. (1987) Detection *in situ* of genomic regulatory elements in *Drosophila*. *Proc. Natl Acad. Sci. USA*, **84,** 9123.
48. Brand, A. H. and Perrimon, N. (1993) Targeted gene expression as a means of altering cell fates and generating dominant phenotypes. *Development*, **118,** 401.
49. Fisher, J. A., Giniger, E., Maniatis, T., and Ptashne, M. (1988) GAL4 activates transcription in *Drosophila*. *Nature*, **332,** 853.
50. Bellen, H. J., Develyn, D., Harvey, M., and Elledge, S. J. (1992) Isolation of temperature-sensitive diphtheria toxins in yeast and their effects in *Drosophila* cells. *Development*, **114,** 787.
51. Moffat, K. G., Gould, J. H., Smith, H. K., and O'Kane, C. J. (1992) Inducible cell ablation in *Drosophila* by cold sensitive ricin A chain. *Development*, **114,** 681.
52. O'Kane, C. J. and Moffat, K. G. (1992) Selective cell ablation and genetic surgery. *Curr. Opin. Genet. Dev.*, **2,** 602.
53. Sentry, J. W., Yang, M.-Y., and Kaiser, K. (1993) Conditional cell ablation in *Drosophila*. *BioEssays*, **15,** 491.
54. Voelker, R. A., Greenleaf, A. L., Gyurkovics, H., Wisely, G. B., Huang, S.-M., and Searles, L. L. (1984) Frequent imprecise excision among reversions of a P element-caused lethal mutation in *Drosophila*. *Genetics*, **107,** 279.
55. Tsubota, S. and Schedl, P. (1986) Hybrid dysgenesis-induced revertants of insertions at the 5′ end of the *rudimentary* gene in *Drosophila melanogaster*: transposon-induced control mutants. *Genetics*, **114,** 165.
56. Salz, H. K., Cline, T. W., and Schedl, P. (1987) Functional changes associated with structural alterations induced by mobilization of a P element inserted in the *sex-lethal* gene in *Drosophila*. *Genetics*, **117,** 221.
57. Kaiser, K. (1990) From gene to phenotype in *Drosophila* and other organisms. *BioEssays*, **12,** 297.
58. Segalat, L., Perichon, R., Bouly, J. P., and Lepesant, J. A. (1992) The *Drosophila pourquoi-pas?/wings-down* zinc finger protein: oocyte nucleus localization and embryonic requirement. *Genes Dev.*, **6,** 1019.
59. Cooley, L., Thompson, D., and Spradling, A. C. (1990) Constructing deletions with defined endpoints in *Drosophila*. *Proc. Natl Acad. Sci. USA*, **87,** 3170.
60. Gloor, G. B., Nassif, N. A., Johnson-Schlitz, D. M., Preston, C. R., and Engels, W. R. (1991) Targeted gene replacement in *Drosophila* via P element-induced gap repair. *Science*, **253,** 1110.

61. Nassif, N. and Engels, W. (1993) DNA homology requirements for mitotic gap repair in *Drosophila*. *Proc. Natl Acad. Sci. USA*, **90,** 1262.
62. Handler, A. M., Gomez, S. P., and Obrochta, D. A. (1993) A functional analysis of the P element gene transfer vector in insects. *Arch. Insect Biochem. Physiol.*, **22,** 373.
63. Kaufman, P. D. and Rio, D. C. (1992) P element transposition *in vitro* proceeds by a cut and paste mechanism and uses GTP as a cofactor. *Cell*, **69,** 27.
64. Calvi, B. R., Hong, T. J., Findley, S. D., and Gelbart, W. M. (1991) Evidence for a common evolutionary origin of inverted repeat transposons in *Drosophila* and plants—hobo, activator, and Tam3. *Cell*, **66,** 465.
65. Yannopoulos, G., Stamatis, N., Monastirioti, M., Hatzopoulous, P., and Louis, C. (1987) *Hobo* is responsible for the induction of hybrid dysgenesis by strains of *Drosophila melanogaster* bearing the male recombination factor 23.5MRF. *Cell*, **49,** 487.
66. Blackman, R. K., Grimaila, R., Koehler, M. M. D., and Gelbart, W. M. (1987) Mobilization of *hobo* elements residing within the decapentaplegic gene complex—suggestion of a new hybrid dysgenesis system in *Drosophila melanogaster*. *Cell*, **49,** 497.
67. Lim, J. K. (1988) Intrachromosomal rearrangements mediated by *hobo* transpositions in *Drosophila melanogaster*. *Proc. Natl Acad. Sci. USA*, **85,** 9153.
68. Blackman, R. K., Koehler, M. M. D., Grimaila, R., and Gelbart, W. M. (1989) Identification of a fully functional *hobo* transposable element and its use in germ-line transformation of *Drosophila*. *EMBO J.*, **8,** 211.
69. Jacobson, J. W., Medhora, M. M., and Hartl, D. L. (1986) Molecular structure of a somatically unstable transposable element in *Drosophila*. *Proc. Natl Acad. Sci. USA*, **83,** 8684.
70. Medhora, M., Maruyama, K., and Hartl, D. L. (1991) Molecular and functional analysis of the *mariner* mutator element *Mos*1 in *Drosophila*. *Genetics*, **128,** 311.
71. Bryan, G., Garza, D., and Hartl, D. (1990) Insertion and excision of the transposable element *mariner* in *Drosophila*. *Genetics*, **125,** 103.
72. Lidholm, D. D., Lohe, A. R., and Hartl, D. L. (1993) The transposable element *mariner* mediates germline transformation in *Drosophila melanogaster*. *Genetics*, **134,** 839.
73. Lidholm, D. A., Gudmundsson, G. H., and Boman, H. G. (1991) A highly repetitive, *mariner*-like element in the genome of *Hyalophora cecropia*. *J. Biol. Chem.*, **266,** 11518.
74. Robertson, H. M. (1993) The *mariner* transposable element is widespread in insects. *Nature*, **362,** 241.
75. Robertson, H. M. (1993) Infiltration of *mariner* elements: reply. *Nature*, **364,** 109.
76. García-Fernandez, J., Marfany, G., and Salo, E. (1993) Infiltration of *mariner* elements. *Nature*, **364,** 109.
77. Sulston, J., Du, Z., Thomas, K., Wilson, R., Hillier, L., Staden, R., Halloran, N., Green, P., Thierrymieg, J., Qiu, L., Dear, S., Coulson, A., Craxton, M., Durbin, R., Berks, M., Metzstein, M., Hawkins, T., Ainscough, R., and Waterston, R. (1992) The *C. elegans* genome sequencing project—a beginning. *Nature*, **356,** 37.
78. Garza, D., Medhora, M., Koga, A., and Hartl, D. L. (1991) Introduction of the transposable element *mariner* into the germline of *Drosophila melanogaster*. *Genetics*, **128,** 303.
79. Franz, G. and Savakis, C. (1991) *Minos*, a new transposable element from *Drosophila hydei* is a member of the Tc1-like family of transposons. *Nucleic Acids Res.*, **19,** 6646.
80. Finnegan, D. J. (1989) The I Factor and I–R hybrid dysgenesis in *Drosophila melano-*

gaster. In *Mobile DNA*. D. E. Berg and M. M. Howe (eds). American Society for Microbiology, Washington, DC, p. 503.
81. Milligan, C. D. and Kaiser, K. (1993) 'Site-selected' mutagenesis of a *Drosophila* gene using the I factor retrotransposon. *Nucleic Acids Res.*, **21,** 1323.
82. Golic, K. G. and Lindquist, S. (1989) The FLP recombinase of yeast catalyzes site-specific recombination in the *Drosophila* genome. *Cell,* **59,** 499.
83. Cox, M. M. (1988) FLP site-specific recombination system of *Saccharomyces cerevisiae*. In *Genetic Recombination*. R. Kucherlapati and G. R. Smith (eds). American Society for Microbiology, Washington, DC, p. 429.
84. Struhl, G. and Basier, K. (1993) Organizing activity of *wingless* protein in *Drosophila*. *Cell,* **72,** 527.
85. Golic, K. G. (1991) Site-specific recombination between homologous chromosomes in *Drosophila*. *Science,* **252,** 958.
86. Cho, T.-B. and Perrimon, N. (1992) Use of a yeast site-specific recombinase to produce female germline chimeras in *Drosophila*. *Genetics,* **131,** 643.
87. Simpson, P. (1993) Flipping fruit-flies: a powerful new technique for generating *Drosophila* mosaics. *Trends Genet.*, **9,** 227.
88. Xu, T. and Rubin, G. M. (1993) Analysis of genetic mosaics in developing and adult *Drosophila* tissues. *Development,* **117,** 1223.
89. Sternberg, N. and Hamiliton, D. (1981) Bacteriophage P1 site-specific recombination between loxp sites. *J. Mol. Biol.*, **150,** 467.
90. Abremski, K., Hoess, R., and Sternberg, N. (1983) Studies on the properties of P1 site-specific recombination—evidence for topologically unlinked products following recombination. *Cell,* **32,** 1301.
91. Abremski, K. and Hoess, R. (1984) Bacteriophage P1 site-specific recombination—purification and properities of the CRE recombinase protein. *J. Biol. Chem.*, **259,** 1509.
92. Fukushige, S. and Sauer, B. (1992) Genomic targeting with a positive-selection lox integration vector allows highly reproducible gene expression in mammalian cells. *Proc. Natl Acad. Sci. USA,* **89,** 7905.
93. Lakso, M., Sauer, B., Mosinger, B., Jr, Lee, E. F., Manning, R. W., Yu, S.-H., Mulder, K. L., and Westphal, H. (1992) Targeted oncogene activation by site-specific recombination in transgenic mice. *Proc. Natl Acad. Sci. USA,* **89,** 6232.
94. Orban, P. C., Chui, D., and Marth, J. D. (1992) Tissue- and site-specific DNA recombination in transgenic mice. *Proc. Natl Acad. Sci. USA,* **89,** 6861.
95. Odell, J., Caimi, P., Sauer, B., and Russell, S. H. (1990) Site-directed recombination in the genome of transgenic tobacco. *Mol. Gen. Genet.*, **223,** 369.
96. Dale, E. M. and Ow, D. W. (1991) Gene transfer with subsequent removal of the selection gene from the host genome. *Proc. Natl Acad. Sci. USA,* **88,** 10558.
97. Moerman, D. G. and Waterston, R. H. (1989) Mobile elements in *Caenorhabditis elegans* and other nematodes. In *Mobile DNA*. D. E. Berg and M. M. Howe (eds). American Society for Microbiology, Washington, DC, p. 537.
98. Harris, L. J., Prasad, S., and Rose, A. M. (1990) Isolation and sequence analysis of *Caenorhabditis briggsae* repetitive elements related to the *Caenorhabditis elegans* transposon Tc1. *J. Mol. Evol.*, **30,** 359.
99. Abad, P. A., Quiles, C., Tares, S., Piotte, C., Castagnone-Sereno, P., Abadon, M., and Dalmasso, A. (1991) Sequences homologous to Tc(s) transposable elements of *Caenorhabditis elegans* are widely distributed in the phylum nematoda. *J. Mol. Evol.*, **33,** 251.

100. Vos, J. C., van Luenen, H. G. A. M., and Plasterk, H. (1993) Characterization of the *Caenorhabditis elegans* Tc1 element transposase *in vivo* and *in vitro*. *Genes Dev.*, **7**, 1244.
101. Plasterk, R. H. A. (1991) The origin of footprints of the Tc1 transposon in *Caenorhabditis elegans*. *EMBO J.*, **10**, 1919.
102. Greenwald, I. (1985) *lin*-12, a nematode homeotic gene, is homologous to a set of mammalian proteins that includes growth factor. *Cell*, **43**, 583.
103. Moerman, D. G., Benian, G. M., and Waterston, R. H. (1986) Molecular cloning of the muscle gene *unc*-22 in *Caenorhabditis elegans* by Tc1 transposon tagging. *Proc. Natl Acad. Sci. USA*, **83**, 2579.
104. Rushforth, A. M., Saari, B., and Anderson, P. (1993) Site-selected insertion of the transposon Tc1 into a *Caenorhabditis elegans* myosin light chain gene. *Mol. Cell. Biol.*, **13**, 902.
105. Andach, Y. and Kohora, Y. (1993) Construction of a Tc1 insertional bank of *C. elegans*. *W.B.G.*, **12**, 21.
106. Zwarl, R. R., Broeks, A., van Meurs, J., Groenen, J. T. M., and Plasterk, R. (1993) Target-selected gene inactivation in *Caenorhabditis elegans*, using a frozen insertion mutant bank. *Proc. Natl Acad. Sci. USA*, **90**, 7431.
107. Plasterk, R. H. A. (1992) Reverse genetics of *Caenorhabditis elegans*. *BioEssays*, **14**, 629.
108. Broeks, A., Zwaal, R., and Plasterk, R. H. A. (1993) Transposon induced deletion mutagenesis, a general method. *W.B.G.*, **12**, 20.
109. Plasterk, R. H. A. and Groenen, J. T. M. (1993) Targeted alterations of the *Caenorhabditis elegans* genome by transgene instructed DNA double strand break repair following Tc1 excision. *EMBO J.*, **11**, 287.
110. van Luenen, H. G. A. M., Colloms, S. D., and Plasterk, R. H. A. (1993) Mobilization of quiet, endogenous Tc3 transposons of *Caenorhabditis elegans* by forced expression of Tc3 transposase. *EMBO J.*, **12**, 2513.
111. Collins, J., Forbes, E., and Anderson, P. (1989) The Tc3 family of transposable genetic elements in *Caenorhabditis elegans*. *Genetics*, **121**, 47.
112. McClintock, B. (1948) Mutable loci in maize. *Carnegie Inst. Wash. Publ.*, **47**, 155.
113. Fedoroff, N. (1989) Maize transposable elements. In *Mobile DNA*. D. E. Berg and M. M. Howe (eds). American Society for Microbiology, Washington, DC, p. 375.
114. Fedoroff, N., Wessler, S., and Shure, M. (1983) Isolation of the controlling maize elements *Ac* and *Ds*. *Cell*, **35**, 243.
115. Kunze, R., Stochaj, U., Laufs, J., and Starlinger, P. (1987) Transcription of transposable element *Activator* (*Ac*) of *Zea mays* L. *EMBO J.*, **6**, 1555.
116. Finnegan, E. J., Taylor, B. H., Dennis, E. S., and Peacock, W. J. (1988) Transcription of the maize transposable element *Ac* in maize seedlings and in transgenic tobacco. *Mol. Gen. Genet.*, **212**, 505.
117. Coupland, G., Baker, B., and Starlinger, P. (1988) Characterization of the maize transposable element *Ac* by internal deletions. *EMBO J.*, **7**, 3653.
118. Kunze, R. and Starlinger, P. (1989) The putative transposase of transposable element *Ac* from *Zea mays* L. interacts with subterminal sequences of *Ac*. *EMBO J.*, **8**, 3177.
119. Coupland, G., Plum, C., Chatterjee, S., Post, A., and Starlinger, P. (1989) Sequences near the termini are required for transposition of the maize transposon *Ac* in transgenic tobacco plants. *Proc. Natl Acad. Sci. USA*, **86**, 9385.
120. Sutton, W. D., Gerlach, W. L., Schwarz, D., and Peacock, W. J. (1984) Molecular analysis of *Ds* controlling element mutations at the *Adh* locus of maize. *Science*, **223**, 1265.

121. Gierl, A. and Saedler, H. (1992) Plant transposable elements and gene tagging. *Plant Mol. Biol.*, **19,** 39.
122. Greenblatt, I. M. and Brink, R. A. (1962) Twin mutations in medium variegated pericarp maize. *Genetics*, **47,** 489.
123. Jones, J. D. G., Carland, F., Lim, E., Ralston, E., and Dooner, H. K. (1990) Preferential transposition of the maize element *Activator* to linked chromosomal locations in tobacco. *Plant Cell*, **2,** 701.
124. Swinburne, J., Balcells, L., Scofield, S. R., Jones, J. D. G., and Coupland, G. (1992) Elevated levels of *Ac* transposase mRNA are associated with high frequencies of *Ds* excision in *Arabidopsis*. *Plant Cell*, **4,** 583.
125. Bancroft, I., Jones, J. D. G., and Dean, C. (1993) Heterologous transposon tagging of the *DRL 1* locus in *Arabidopsis*. *Plant Cell*, **5,** 631.
126. Long, D., Martin, M., Sundberg, E., Swinburne, J., Poangsomlee, P., and Coupland, G. (1993) The maize transposable element system *Ac/Ds* as a mutagen in *Arabidopsis*: identification of an *albino* mutation induced by *Ds* insertion. *Proc. Natl Acad. Sci. USA*, **90,** 10370.
127. Chuck, G., Robbins, T., Nijar, C., Ralston, E., Courtney-Gutterson, N., and Dooner, H. K. (1993) Tagging and cloning of a petunia flower color gene with the maize transposable element *Activator*. *Plant Cell*, **5,** 371.
128. Fedoroff, N. V. and Smith, D. L. (1993) A versatile system for detecting transposition in *Arabidopsis*. *Plant Cell*, **3,** 273.
129. Gierl, A., Saedler, H., and Pertson, P. A. (1989) Maize transposable elements. *Annu. Rev. Genet.*, **23,** 71.
130. Masson, P., Rutherford, G., Banks, J. A., and Fedoroff, N. (1989) Essential large transcripts of the maize *Spm* transposable element are generated by alternative splicing. *Cell*, **58,** 755.
131. Frey, M., Reinicke, J., Grant, S., Saedler, H., and Gierl, A. (1990) Excision of En Spm transposable element of *Zea mays* requires two element-encoded proteins. *EMBO J.*, **9,** 4037.
132. Masson, P., Strem, M., and Fedoroff, N. (1991) The tnpA and tnpD gene products of the Spm element are required for transposition in tobacco. *Plant Cell*, **33,** 78.
133. Gierl, A., Lütticke, S., and Saedler, H. (1988) *TnpA* product encoded by the transposable element En-1 of *Zea mays* is a DNA binding protein. *EMBO J.*, **7,** 4045.
134. Masson, P. and Fedoroff, N. (1989) Mobility of the maize Suppressor–mutator element in transgenic tobacco cells. *Proc. Natl Acad. Sci. USA*, **86,** 2219.
135. Aarts, M. G. M., Dirkse, W. G., Stiekema, W. J., and Pereira, A. (1993) Transposon tagging of a male sterility gene in *Arabidopsis*. *Nature*, **363,** 715.
136. Coen, E. S., Robbins, T. P., Almeida, J., Hudson, A., and Carpenter, R. (1989) Consequences and mechanisms of transposition in *Antirrhinum majus*. In *Mobile DNA*. D. E. Berg and M. M. Howe (eds). American Society for Microbiology, Washington, DC, p. 41.
137. Nacken, W. K. F., Piotrowiak, R., Saedler, H., and Sommer, H. (1991) The transposable element Tam1 from *Antirrhinum majus* shows structural homology to the maize transposon En/Spm and has no sequence specificity of insertion. *Mol. Gen. Genet.*, **228,** 201.
138. Hehl, R., Nacken, W. K. F., Krause, A., Saedler, H., and Sommer, H. (1991) Structural analysis of Tam3, a transposable element from *Antirrhinum majus*, reveals homologies to the Ac element from maize. *Plant Mol. Biol.*, **16,** 369.

139. Haring, M. A., Gao, J., Volbeda, T., Rommens, C. M. T., Nijkamp, H. J. J., and Hille, J. (1989) A comparative study of *Tam3* and *Ac* transposition in transgenic tobacco and petunia plants. *Plant Mol. Biol.*, **13,** 189.
140. Martin, C., Prescott, A., Lister, C., and Mackay, S. (1989) Activity of transposon *Tam3* in *Antirrhinum* and tobacco: possible role of DNA methylation. *EMBO J.*, **8,** 997.
141. Jacobs, E., Dewerchin, M., and Boeke, J. D. (1988) Retrovirus-like vectors for *Saccharomyces cerevisiae*—integration of foreign genes controlled by efficient promoters into yeast chromosomal DNA. *Gene*, **67,** 259.
142. Lineruth, K., Duncanson, A., Kaiser, K., O'Dell, K., and Davis, T. (1992) The isolation and characterisation of P element insertions into G protein genes. *Biochem. Soc. Trans.*, **20,** 261S.
143. Pereira, A., Doshen, J., Tanaka, E., and Goldstein, L. S. B. (1992) Genetic analysis of a *Drosophila* microtubule-associated protein. *J. Cell Biol.*, **116,** 377.
144. Segalat, L., Perichon, R., Bouly, J. P., and Lepesant, J. A. (1992) The *Drosophila pourquoi-pas?/wings-down* zinc finger protein: oocyte nucleus localisation and embryonic requirement. *Genes Dev.*, **6,** 1019.
145. Laufs, J., Wirtz, U., Kammann, N., Matzeit, V., Schaefer, S., Schell, J., Czernilofsky, A. P., Baker, B., and Grienenborn, B. (1990) Wheat dwarf virus *Ac/Ds* vectors—expression and excision of transposable elements introduced into various cereals by a viral replicon. *Proc. Natl Acad. Sci. USA*, **87,** 7752.
146. Baker, B., Schell, J., Lorz, H., and Fedoroff, N. (1986) Transposition of the maize controlling element 'Activator' in tobacco. *Proc. Natl Acad. Sci. USA*, **83,** 4844.
147. Marion Poll, A., Marin, E., Bonnefoy, N., and Pautot, V. (1993) Transposition of the maize autonomous element *Activator* in transgenic *Nicotiana plumbaginifolia* plants. *Mol. Gen. Genet.*, **238,** 209.
148. Yoder, J. L., Palys, J., Alpert, K., and Lassner, M. (1988) *Ac* transposition in transgenic tomato plants. *Mol. Gen. Genet.*, **213,** 291.
149. Knapp, S., Coupland, G., Uhrig, H., Starlinger, P., and Salamini, F. (1988) Transposition of the maize transposable element *Ac* in *Solanum tuberosum*. *Mol. Gen. Genet.*, **213,** 285.
150. Van Sluys, M. A., Tempe, J., and Fedoroff, N. (1987) Studies on the introduction and mobility of the maize *Activator* element in *Arabidopsis thaliana* and *Daucus carota*. *EMBO J.*, **6,** 3881.
151. Zhou, J. H. and Atherly, A. G. (1990) *In situ* detection of transposition of the maize controlling element (*Ac*) in transgenic soybean tissue. *Plant Cell Rep.*, **8,** 542.
152. Finnegan, E. J., Lawrence, G. J., Dennis, E. S., and Ellis, J. G. (1993) Behaviour of modified *Ac* elements in flax callus and regenerated plants. *Plant Mol. Biol.*, **22,** 625.

5 | Topological selectivity in site-specific recombination

W. MARSHALL STARK and MARTIN R. BOOCOCK

1. Introduction

In conservative site-specific recombination, DNA molecules are broken in both strands at two separate predetermined points, and the ends are rejoined to new partners, without any synthesis or degradation of the DNA (Fig. 1a). The reactions are catalysed by specialized recombinase proteins and may involve other protein accessory factors. Many site-specific recombination systems have been identified in prokaryotes and eukaryotes, and several systems have been reconstituted *in vitro*. In studying these systems, one of the main aims has been to attempt to understand the mechanisms by which selectivity for recombination between sites in particular relationships is achieved. It has become apparent that DNA topology plays a crucial role in this selectivity. *In vivo*, site-specific recombination has a variety of functions, including switching of gene expression, control of plasmid copy number, resolution of plasmid multimers to monomers, bacteriophage integration and excision, and transposon co-integrate resolution. Selectivity is important to ensure that only suitable pairs of sites recombine, since uncontrolled recombination would lead to loss of function, and might create undesirable genetic

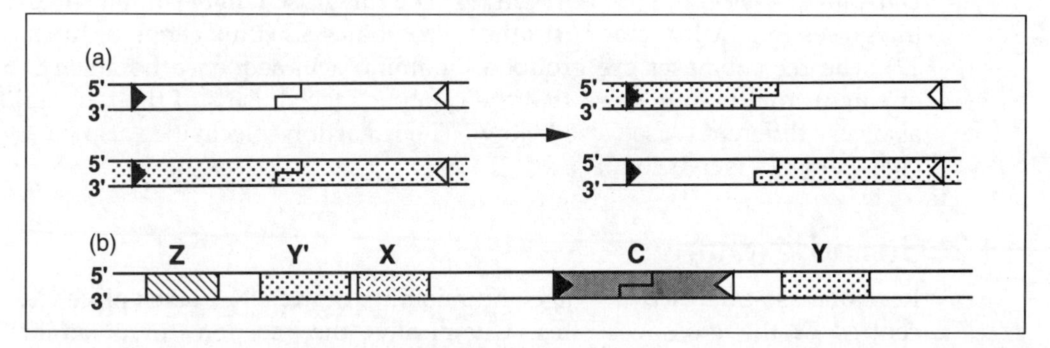

Fig. 1 (a) Site-specific recombination. The DNA is broken at two sites, and the ends are exchanged, with no synthesis or degradation of DNA. (The DNA is represented as a ribbon.) (b) A typical recombination site. The strand cutting and rejoining takes place within a short sequence (C) with some inverted repeat symmetry, that binds recombinase. Nearby there may be other binding sites, for recombinase or accessory proteins (X, Y, Y′, and Z). Actual sites are all variations on this theme.

rearrangements. References 1–4 are recent reviews on general aspects of the mechanisms of site-specific recombination reactions.

There are many parallels between the mechanisms of site-specific recombination and those of genetic transposition, the main focus of this volume. In both processes, two or more segments of DNA are brought together by means of proteins that interact with specific sites, and then a series of DNA strand cleavages and religations takes place. Transposition reactions also show topological selectivity, and it is very likely that there are similarities in the mechanisms used by the two types of system to achieve this selectivity. See other chapters in this book, and reference 5, a comprehensive review on the mechanism of phage Mu transposition, for details.

1.1 Sites and recombinases

The minimal sites required for recombination vary widely in length among systems. The length differences reflect heterogeneity in the requirements for binding sites for proteins involved in the reaction. Some systems have only a pair of identical short (~ 30 bp) sites, which usually have some inverted repeat sequence symmetry. Each site might bind two subunits of recombinase (6). Larger sites seem to be made up from an essentially similar core or crossover site, surrounded by or adjacent to additional accessory binding sites, for recombinase or other proteins (Fig. 1b). A well-known example of a complex site is phage λ *att*P (240 bp) which has a crossover site that binds Int recombinase, flanked by sequences that bind a different domain of Int, and IHF, Xis, and FIS proteins. *att*B, in contrast, is a simple crossover site (7). Parts of recombination sites may be adapted for bending or twisting the DNA into conformations appropriate for their reaction (7–10).

The recombinases are the proteins believed to be the protagonists in the chemical catalytic steps in recombination. There are two distinct families of recombinases, conveniently referred to by the names of two of their well-known members. The resolvase family contains many transposon resolvases (e.g. that of Tn3) and the DNA invertases Gin, Hin, and Cin (9). The integrase family contains many phage integrases (e.g. λ Int) amongst other recombinases with a range of functions (11, 12). The recombinases are grouped by amino acid sequence homology, but it is apparent from *in vitro* work that the catalytic mechanisms of the two families are also very different (3; see also below). Their reaction selectivities also have familial characteristics (see below).

1.2 Strand exchange

Recombinase-catalysed cleavage and rejoining of the DNA takes place close to the centres of the crossover sites. Chemically, the reactions involved are transesterifications; first, a phosphate group of the DNA backbone is transferred to a recombinase hydroxyl group, making a protein–DNA covalent phosphodiester linkage, and breaking the DNA chain to release a deoxyribose hydroxyl group; then the phosphate is transferred to a new deoxyribose hydroxyl, re-forming the

DNA backbone. The top and bottom strand cleavages are at specific phosphodiesters, and are staggered by 2 bp (resolvase family) or 6–8 bp (integrase family) (Fig. 2a). Resolvase family enzymes apparently cut the DNA at the paired sites in all four strands, before exchanging the ends by a mechanism that is consistent with a 'simple rotation' model (Fig. 2b). Integrase-type enzymes exchange strands in pairs; exchange of one pair creates a four-way junction (Holliday junction) which branch-migrates through the short overlap region between the two points of exchange before being resolved to recombinant by a second pair of strand exchanges. Figure 2c illustrates a model for the movements of DNA and protein subunits during the integrase-type reaction, that is consistent with the available evidence (3, 13). It is important for normal strand exchange in both types of system that the overlap regions in a pair of recombining sites are identical in sequence, because if they are not, the recombinant sites would contain mismatched base-pairs (see also below).

1.3 The complete reaction

A site-specific recombination reaction can be divided conceptually into four stages (Fig. 3), each of which may consist of many individual steps. First, the recombinase, and in some cases accessory proteins, bind to one or both sites (binding). Secondly, the two sites that are to recombine come together to form a protein–DNA complex that is competent for recombination (synapsis). Thirdly, the DNA strands are broken, and rejoined to new partners (strand exchange). Finally, the complex containing the paired recombinant DNA sites dissociates ('desynapsis'). *In vitro* the recombinant sites are likely to remain in association with recombinase until the reaction is stopped (e.g. by denaturation of the protein), and until then may be available for further rounds of recombination. The four stages might not be completely separate; for example, it might be possible or even necessary to recruit protein subunits to the synaptic complex after the sites have been brought together.

Our information on the rates and reversibility of these stages for any system is at best scanty. In principle, a mechanism for achieving selectivity could act at any stage, or at more than one stage.

1.4 Outcomes of recombination reactions

Consider a circular DNA molecule containing two recombination sites. If the DNA is broken at the two sites, four ends are generated. The ends can be rejoined in three different ways (Fig. 4).

(1) the ends are rejoined as before;

(2) the ends are rejoined to make two smaller circles: this is called resolution, deletion, or excision;

(3) the ends are rejoined to make a single circle, but such that the relative arrangement of the two pieces of DNA has been 'flipped'; this is called inversion.

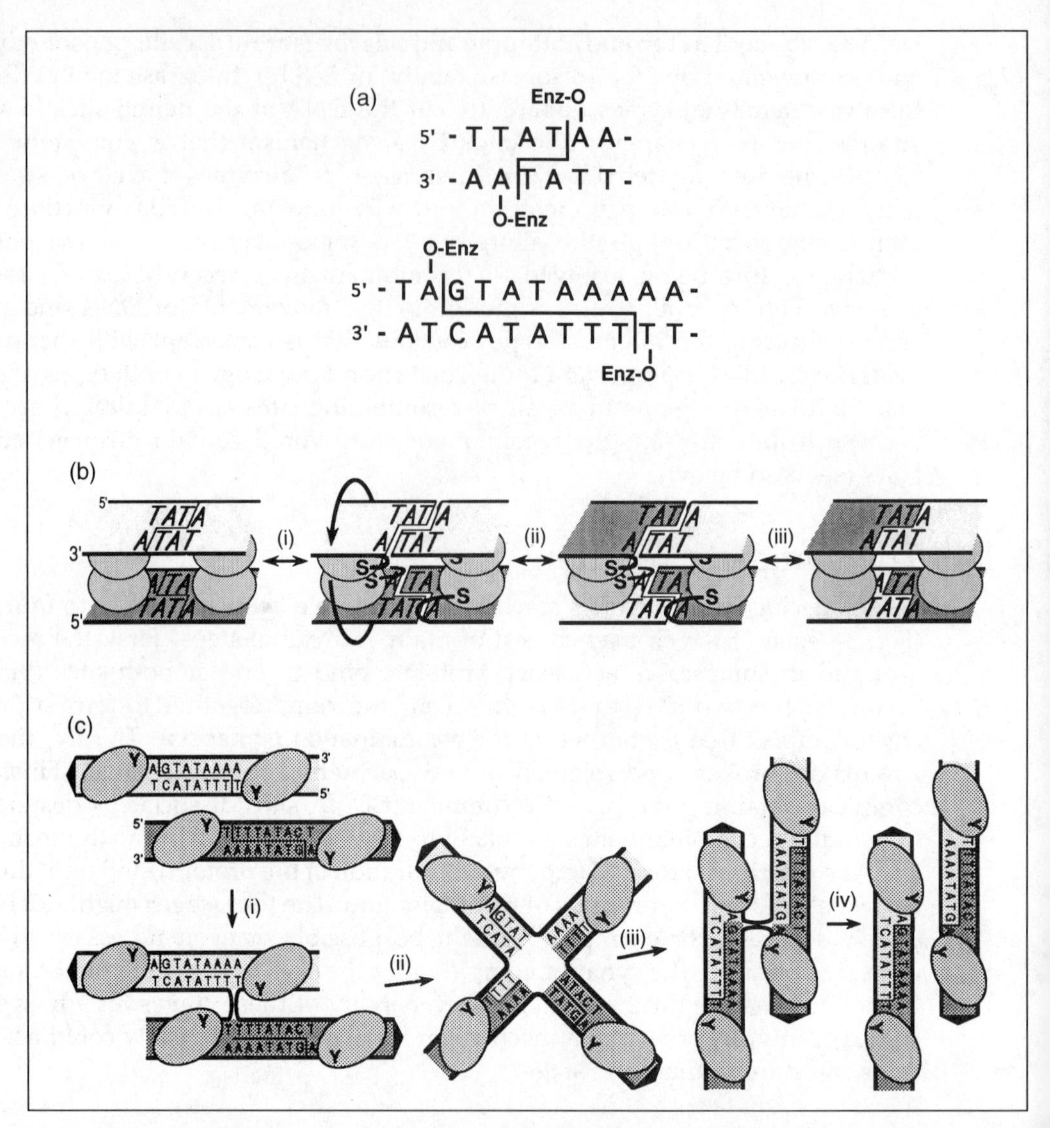

Fig. 2 (a) How recombinases cleave the DNA at the recombination site. Top, resolvase family enzymes. The two breaks are staggered by 2 bp, and the protein (Enz-O-) becomes covalently linked to the 5′ ends of the DNA. The sequence illustrated is from the Tn*3* recombination site *res*. Bottom, integrase family enzymes. The breaks are staggered by 6–8 bp, and the recombinase becomes linked to the 3′ end of the DNA at the breaks. The sequence illustrated is from the phage λ *att* sites. The top and bottom strand breaks are not made simultaneously (see panel (c)). (b) Model for resolvase family strand exchange (3, 13). (i) cleavage; (ii) rotation; (iii) religation. The double helical DNA is represented as a flat ribbon, and the recombinase subunits as shaded blobs. S is the nucleophilic serine residue of the recombinase. The sequence shown in the figure is that of the Tn*3 res* crossover site. (c) Model for integrase family strand exchange (3, 13). The conventions used are as in panel (b). Y is the recombinase's nucleophilic tyrosine residue. The sites are aligned in antiparallel (black arrowheads). (i) First pair of strand exchanges to form a Holliday junction; (ii) and (iii) branch migration and isomerization of the Holliday junction; (iv) second pair of strand exchanges to resolve the Holliday junction. In real, double-helical DNA, steps (ii) and (iii) would also require rotation of the DNA arms about their long axes (see ref. 13). The sequence shown in the figure is that of the λ *att* sites.

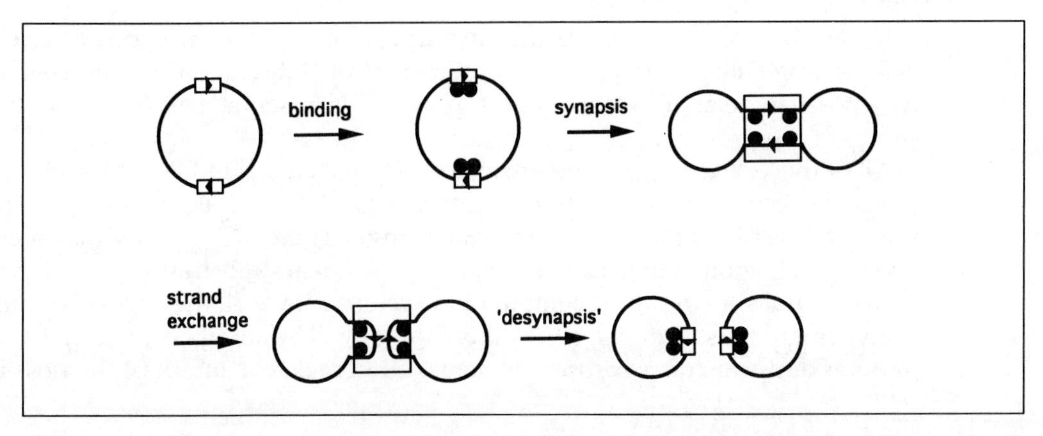

Fig. 3 The steps involved in a site-specific recombination reaction. The sites are represented by arrowheads within rectangles, and proteins (including recombinase) by black circles.

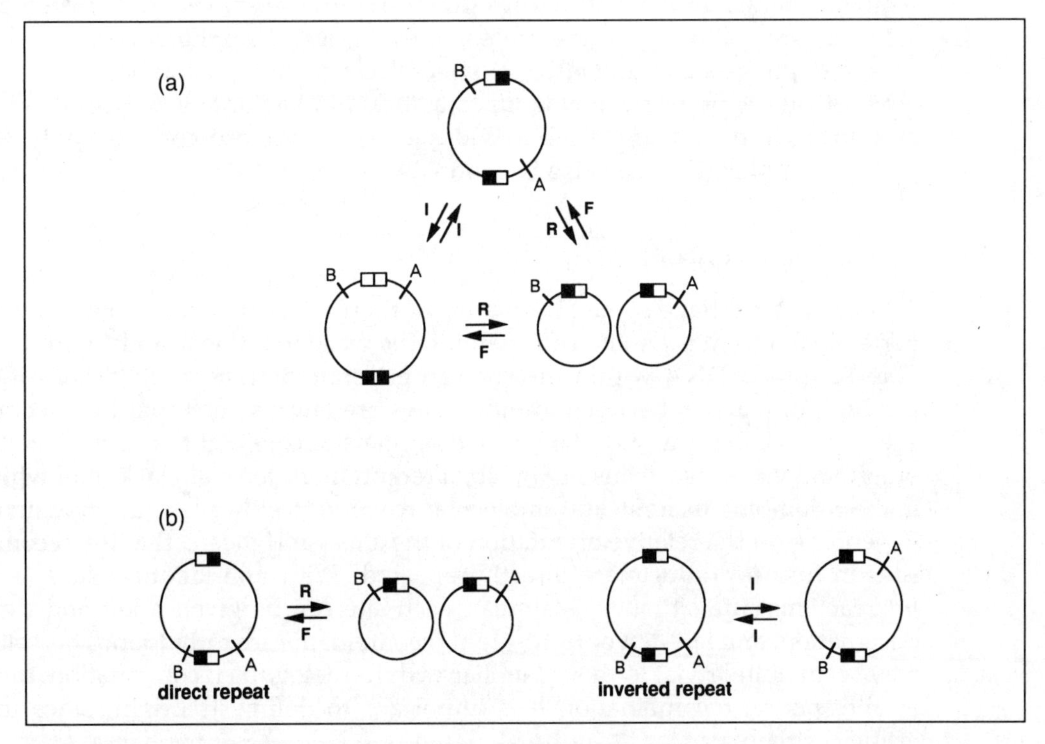

Fig. 4 (a) Possible outcomes from site-specific recombination of circular molecules. **R**, resolution; **F**, fusion; **I**, inversion. (b) Possible outcomes when left and right half-sites must be rejoined to form similar asymmetric sites (see text).

If the two sites are on separate circular molecules, the ends can be rejoined to make a single larger circle. These are 'fusion' or 'integration' reactions. There are two possible relative arrangements of the DNA sequences from the substrate circles in the product.

All of these reactions can be accompanied by changes in DNA topology, i.e. the molecules can become knotted or catenated, and the linking number (the total number of turns of the DNA helix) can change. There are obvious analogies to the above possible outcomes in non-circular (e.g. linear) substrates.

No natural site-specific recombination system allows all these possible outcomes from a single substrate. The rest of this chapter will be about how the interactions of proteins and substrates restrict the actual outcomes to a subset of the possibilities.

2. Selectivities observed in natural systems

2.1 Type of site

Recombinases normally only act on their cognate sites, by recognizing specific sequence motifs and site structures. Many natural reactions are between pairs of identical sites, although there are some examples of reactions between different sites, e.g. phage λ *att*P and *att*B (7). This selectivity for reaction between particular types of sites is probably analogous to a 'lock and key' interaction; each site binds proteins to form a complex that is designed to synapse correctly only with its partner complex (7, 8; see also Section 2.2).

2.2 Left–right selectivity

All natural sites have some asymmetry in their DNA sequences, i.e. they are not perfect palindromes. A site can therefore be assigned a 'left' and 'right' end, and cleavage of the DNA within the site can generate 'left' and 'right' half-sites. The products of reaction between identical sites are always such that the recombinant sites are also identical, i.e. the left half of one site is joined to the right half of the other, and vice versa. Thus in Fig. 4b, a requirement for half black, half white sites in the recombinants limits intramolecular reaction to either resolution or inversion, depending on the relative orientation of the sites, and means that the recombinant sites in fusion products are directly repeated. With non-identical sites (e.g. in λ Int reactions) the situation is similar; each site can be given a left and right end designation, and left–left or right–right recombinants are forbidden. This selectivity may seem quite trivial to those familiar with homologous recombination; however, in site-specific recombination it is important to define its origin, since in some artificial circumstances it can break down (see below).

In many cases the asymmetric properties of the site are determined by asymmetry in sequence between the points of strand exchange on the top and bottom strands. As stated above, these points are staggered by between 2 and 8 bp. If the sequence in this 'overlap' region is asymmetric, left–left or right–right recombinants would

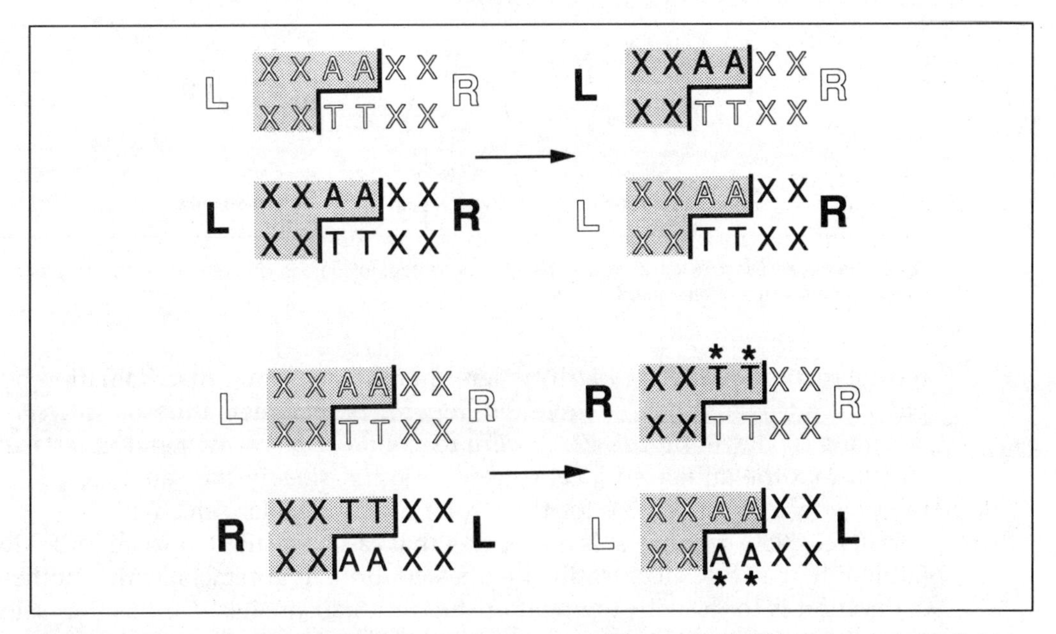

Fig. 5 The importance of homology in the overlap sequence. Top, recombination to create correctly base-paired recombinant sites. Bottom, 'incorrect' left–left and right–right recombinants have mismatched base-pairs (asterisks) in the overlap.

contain mismatched base-pairs (Fig. 5). Recombinases can detect this potential non-homology and maintain the sites in a matched non-recombinant configuration, either by failing to catalyse any strand exchanges, or by reversing any strand exchange steps that have occurred, or by catalysing a second round of recombination (14–21).

For some sites left–right asymmetry is primarily determined by asymmetry in their interaction with recombinase and/or other accessory proteins. For example, the Tn*3* recombination site *res* has a symmetric 2 bp overlap (ApT) but has accessory recombinase-binding subsites only on one side (to the right) of the crossover site (see Section 3.2). It has been shown that the crossover site alone does not behave asymmetrically, but that the complete *res* site's left–right selectivity is determined by the positioning of the accessory subsites (22).

2.3 Connection selectivity

Some reactions are only efficient when the sites are interconnected by DNA in a particular way. There are normally only three possibilities; sites can be directly repeated or inverted, on the same DNA molecule, or they can be on different molecules (Fig. 6). Transposon resolvases (e.g. Tn*3* resolvase) recombine only sites on the same molecule, in direct repeat, whereas the related invertases (e.g. Gin) recombine only sites in inverted repeat (reviewed in ref. 9). This strict selectivity is very different from the properties of some other systems (e.g. Int-catalysed

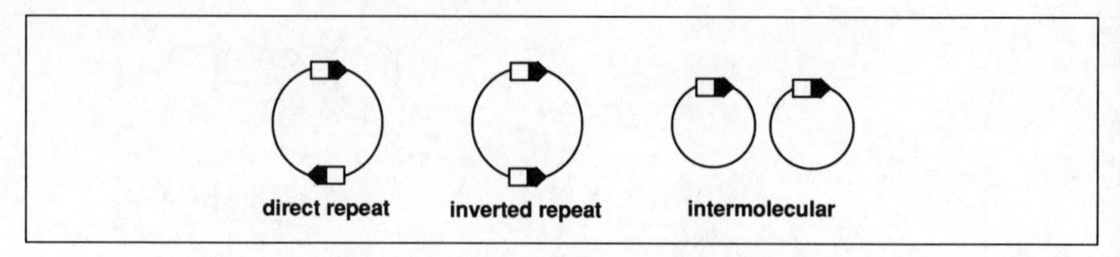

Fig. 6 The three ways in which sites may be interconnected. In some *in vitro* reactions, the circular substrates may also be knotted or catenated (see text).

recombination of *att*P and *att*B) where there is no strong discrimination between pairs of sites in direct or inverted repeat, or between intra-or intermolecular reactions (reviewed in ref. 7). *In vitro* reactions of resolvases and invertases have shown that the enhanced local concentration of sites in the same molecule is not the only factor responsible for the strong intramolecular selectivity.

Whereas 'left–rightness' is a property that can be attributed to an individual site (although it is not necessarily manifested until it interacts with another site), connection is to do with the relationship of a pair of sites. Connection selectivity in site-specific recombination is intimately related to and a consequence of topological selectivity, which is discussed in detail below. The discussion focuses on the resolvases and invertases, the systems that exhibit connection/topological selectivity to the greatest degree.

3. Topological selectivity

3.1 DNA topology and site-specific recombination

DNA molecules with no free ends (e.g. circles) are topologically non-trivial. The topology of such molecules can for practical purposes be partitioned into two types, one being related to the interlinking of the two strands in double-helical DNA, and manifesting itself in some circumstances as supercoiling, and the other being related to the knotting or catenation of the entire DNA double helix. Both types of topology are important in site-specific recombination. For an introduction to DNA topology, see refs 4 and 23.

Here we will be considering the effect of DNA topology mainly in circular molecules. It should be noted that an analogous situation is found wherever two parts of a DNA molecule are anchored, either together to form a loop (e.g. by synapsis of DNA-bound proteins) or individually onto an immobile support (e.g. the nuclear matrix; 24). Most *in vitro* reactions are performed on supercoiled circular DNA molecules, which are analogues of the natural recombination substrates. The topological properties of the substrate can affect all stages of the reaction. However, the topology of the molecule as a whole can only be changed by breaking and re-forming covalent bonds, for example at strand exchange in recombination.

3.2 Recombination product topology

3.2.1 The observations

Tn*3* and γδ resolvases are closely related recombinases that act on sites called *res*. *res* is 114 bp long, and has three imperfect inverted repeat subsites, each of which is thought to bind a resolvase dimer. The cleavage and rejoining reactions take place at the centre of subsite I, as illustrated in Fig. 7 (reviewed in refs 9 and 25).

Early studies on resolvase *in vitro* reactions revealed two curious and apparently unrelated phenomena. First, only directly repeated sites on the same supercoiled molecule would recombine. Prohibition of inverted repeat or intermolecular reactions was essentially absolute. Second, the two circles produced by resolution of the substrate were not separate from each other, but were topologically linked as a simple catenane (26, 27). It was shown subsequently that the product is in fact always one of the two possible simple catenanes, designated (−2) (Fig. 7) (28). These two reaction properties are maintained even when the *res* sites are separated by several kilobases of DNA. The DNA invertases Gin, Hin, and Cin are homologous in amino acid sequence to resolvase, but they only recombine inverted sites, and the inverted products are exclusively unknotted circles (9, 29–32). There is, however, an important difference between the Gin/Hin and resolvase systems: whereas resolvase does not act at all on inverted sites, the invertases do catalyse strand exchange in directly repeated substrates, but recombinant products are not

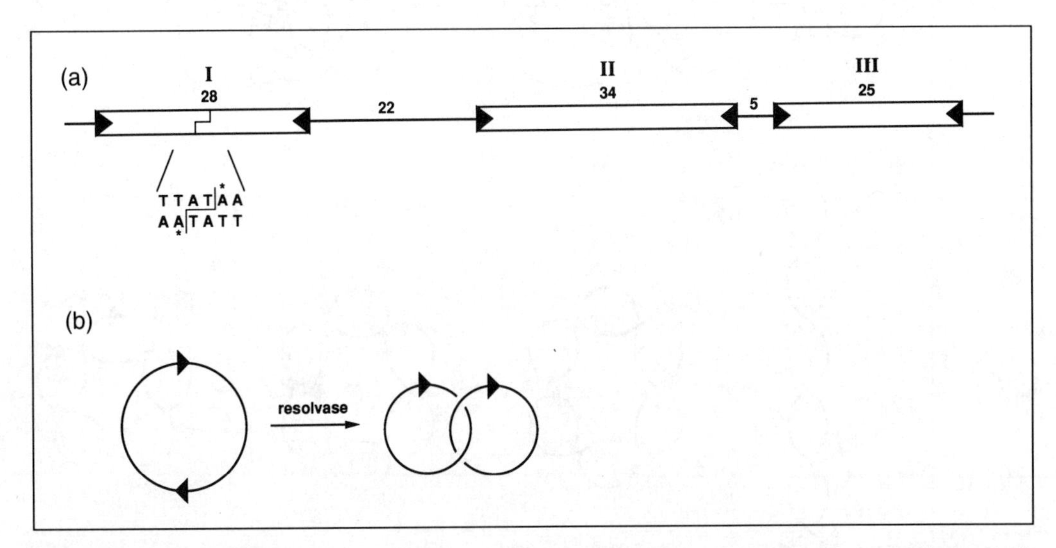

Fig. 7 (a) Tn*3* recombination site, *res*. The three resolvase-binding subsites are shown as boxes, with arrowheads to indicate the imperfect inverted repeats at their outer ends. The small numbers denote the lengths of the DNA segments in base pairs. The sequence at the centre of subsite I, where the DNA is broken and rejoined, is shown (see Fig. 2a). (b) The reaction catalysed *in vitro* by Tn*3* resolvase. A circular (supercoiled) molecule with two directly repeated *res* sites (arrowheads) is resolved into two smaller circles which are simply catenated.

formed (see Section 4.3). The invertases do not require accessory binding sites as in *res*, but do require an enhancer element within the substrate molecule, and a bacterial protein, FIS, that binds to the enhancer (9).

To appreciate how remarkable this selectivity is, it is instructive to consider a tangled pile of string, as a naive analogy of a long DNA molecule (Fig. 8). By simply picking out two 'sites' and bringing them together, there is no obvious way of determining whether they are in direct or inverted repeat, or even if they are on the same piece of string. Also, it is easy to show experimentally that cutting the string at these two points and 'recombining' the ends creates a tangle of indeterminate topology. Try it!

Of course, in reality the structure of supercoiled DNA is rather more ordered. Supercoiled plasmids, *in vitro* and *in vivo*, are thought to be plectonemically wound, as cartooned in Fig. 8b. DNA in this form has been observed by electron microscopy (e.g. ref. 33) and the conformation has been inferred from other physical data (34). It is also consistent with topological data from experiments with topoisomerases and site-specific recombination reactions. Analysis of the products of reactions catalysed by λ Int indicated that their topological form was as would be expected if synapsis of the *att* sites was by 'random collision' within plectonemic

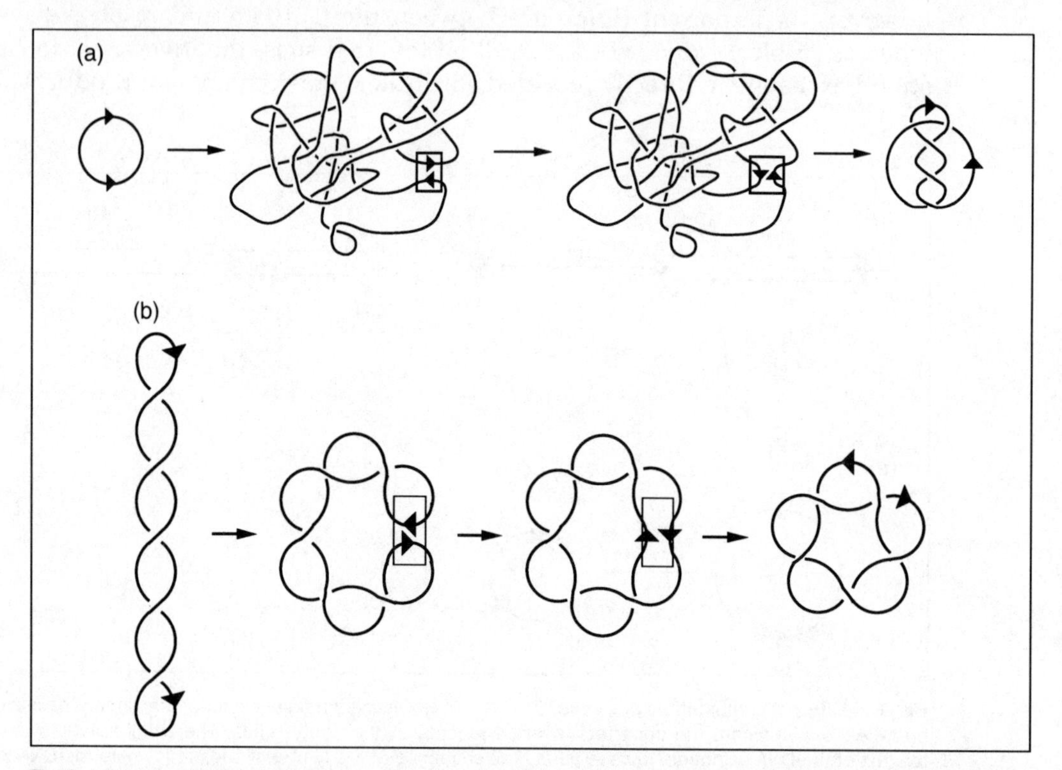

Fig. 8 (a) Recombination between two sites in a randomly tangled circle. In this case the (inversion) product turns out to be a five-noded knot. (b) Recombination between two sites in a plectonemically supercoiled circle. In this case five interdomainal supercoil nodes are trapped as knot nodes in the inversion product.

plasmid molecules (35; Fig. 8b). Also, Int does not distinguish in reactivity between sites in direct or inverted repeat, as would be expected if synapsis was not ordered (7). Similar results were obtained with the FLP and Cre-catalysed systems (Section 3.2.4).

3.2.2 Analysis of recombination topology

Recombination product topology is a consequence of the way that the recombining pair of sites is brought together in the synapse, and the way that the strands are exchanged.

To analyse topological changes during recombination, strand exchange can be imagined to occur within a two-dimensional box containing the aligned crossover sites (Fig. 9). The alignment of the crossover sites is defined as either parallel or antiparallel, with regard to the overall left–right asymmetry of the sites (Section 2.2). The synapse topology is then a description of the entanglement of the DNA outside the box. The strand exchange topology describes the changes that occur inside the box on strand exchange. The strand exchange topology is likely to be fixed for any particular reaction, since it is a representation of the geometrical changes to the DNA within a catalytic protein–DNA complex with a specific structure. On the other hand, synapse topology might be expected to be variable (but see below).

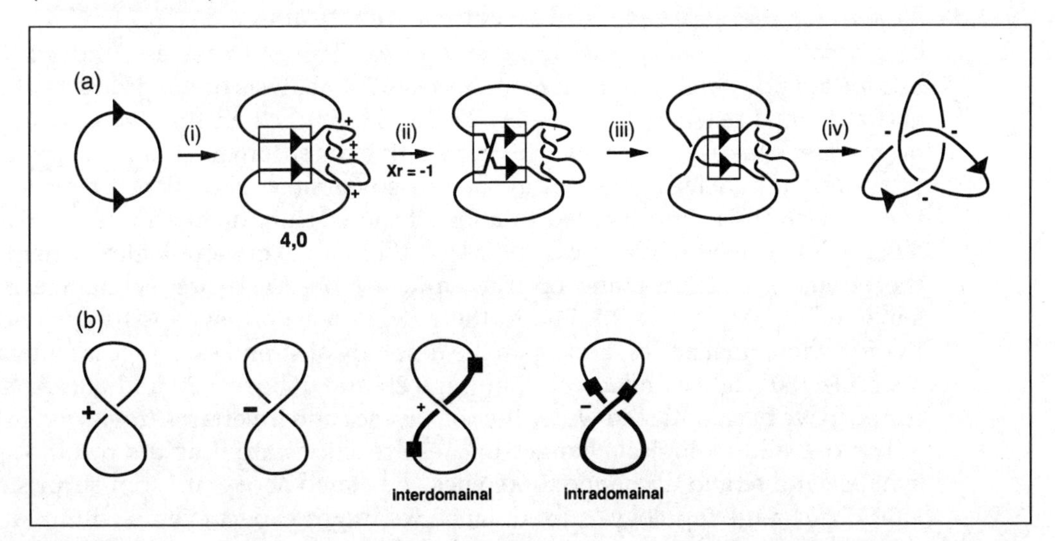

Fig. 9 Topology of a hypothetical recombination reaction. (i) The two inverted repeat sites come together, and are aligned in parallel inside the rectangular box; four (+) nodes are trapped in the DNA outside the synapse (i.e. $S = 4,0$; ref. 13). (ii) Recombinase catalyses strand exchanges in the synapse, introducing a (−) crossing in the diagram (i.e. $Xr = -1$; ref. 13). The geometrical changes that create this node might be entirely within the synaptic complex, or (iii) the node might be 'released' into the external DNA. (iv) Finally the sites dissociate, and the DNA is free to rearrange; in this case the product is a knot ($Kn = -3$; refs 13, 23, 36). (b) Topological nodes in DNA. The conventional signs for nodes are as shown; note that the single line represents the axis of double-stranded DNA. (The conventions for crossings of single strands in the double helix are different; 23.) In the two figures on the right, the domains between two recombination sites (black boxes) are represented as thick and thin lines.

Various approaches have been taken to describe these features mathematically (see, for example, refs 13, 23, 36, and 37). The term node is generally used to describe a point in the two-dimensional representation where two DNA duplex segments cross. Nodes are signed (+) or (−); the convention is illustrated in Fig. 9b. The two crossover sites divide a topologically closed molecule (e.g. a circle) into two domains. Nodes can then be intradomainal or interdomainal, according to whether the crossing DNA segments are from one domain or both domains. The synapse topology (S, or ${}^{ter}Wr^{s}$) can be described in terms of the minimum number of nodes that it contains. The strand exchange topology is usually described by two terms, one representing changes in the twist of the DNA (Xtw, or ${}^{tra}Me$) and the other representing the changes in the way the DNA duplexes are arranged within the box (Xr, or ${}^{ter}Me$).

These methods are very useful for analysis of site-specific recombination reactions, but great care must be taken when using them to infer details of mechanism and synapse geometry from topological data. See, for example, ref. 13 for an illustration of how quite different combinations of synapse and strand exchange topologies might give the same end result. In particular, the alignment of the sites in the box in diagrams as described above does not necessarily imply a similar alignment in the real synapse.

3.2.3 Topology of resolvase and invertase reactions

The topological changes associated with the Tn*3* *res*/resolvase and *gix*/Gin reactions are now well characterized, as a result of analysis of the products of normal and abnormal reactions (13, 19–21, 28–30, 32, 38). When the two crossover sites of *res* are synapsed, exactly three negative interdomainal nodes are trapped (assuming a parallel alignment at the crossover sites; Fig. 10a). Strand exchange is equivalent to a right-handed rotation of the DNA ends in an intermediate with all strands cleaved ($Xr = +1$, $Xtw = +1$) (Fig. 2b). This topological scheme gives the observed (−2) catenane product, and the observed precise change in DNA supercoiling ($\Delta Lk = +4$; 13, 71). In the *gix*/Gin reaction, synapsis traps two negative interdomainal nodes, and again the topology of strand exchange is equivalent to a simple 180° right-handed rotation (Figs 2b and 10b; refs 29 and 30). Analogous results have been obtained with other resolvases and invertases (reviewed in ref. 9).

The overall topological changes in these reactions are thus the result of precise synapse and strand exchange topologies. As stated above, it is not surprising that strand exchange topology is fixed, since we might expect events within an elaborate catalytic complex of protein and DNA to be tightly controlled. But there remains the problem of how a unique synapse topology is chosen (see Section 4).

3.2.4 Integrase family systems

Three integrase family recombinases have been extensively analysed *in vitro* (λ Int, Cre, and FLP; refs 7, 39, 40). A major difference from the resolvases and invertases is that in reactions with these enzymes, mixtures of product topologies can be obtained. The products are consistent with variable synapse topology as a result

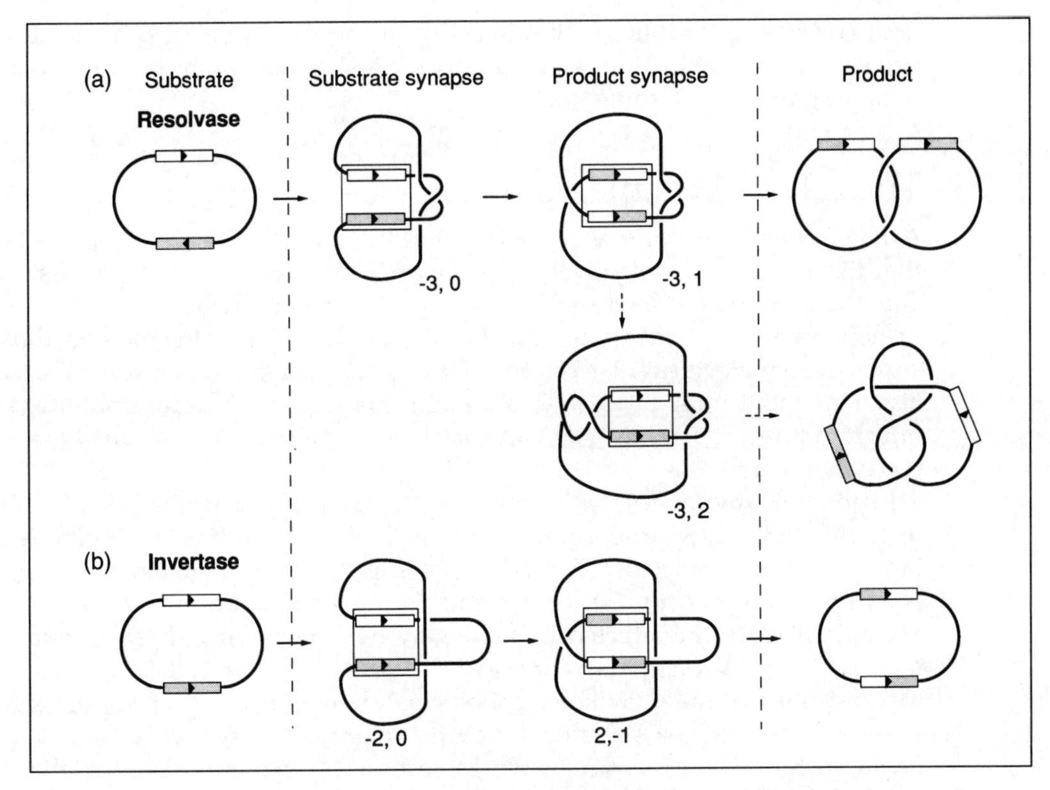

Fig. 10 Topology of recombination by Tn*3* resolvase (top) and Gin invertase (bottom). The boxes with arrows represent the crossover sites (subsite I for *res*). The substrates are normally supercoiled. For both resolvase and invertase reactions, the product molecule has four fewer supercoils than the substrate (i.e. $\Delta Lk = +4$). A double round of recombination reactions by resolvase (dotted arrow pathway) gives a knotted product (13, 20, 29, 30). The numbers on the figure are the synapse topologies, *S* (ref. 13).

of 'random collision' of sites in plectonemic supercoiled DNA (see Fig. 8b). However, for all these systems, the extent of formation of 'random collision' products varies depending on the reaction conditions, substrate structure, and site structure. In some circumstances the products may be predominantly or completely of one topological form, e.g. free circles from *lox*/Cre resolution (39) and from λ Int-catalysed Xis-independent excision (41). Also, mutations in the recombinase can alter product topological complexity (42). Nevertheless, these systems can be described as 'topologically unconstrained'. All three recombinases will recombine sites on the same or different DNA circles, and in direct or inverted repeat, in strong contrast to the strict selectivity of the resolvases and invertases.

The linkage changes associated with reactions catalysed by λ Int, FLP, and Cre have been determined (43–45). As with the resolvase family reactions, discrete values of ΔLk are observed, implying that the strand exchange mechanisms have precise geometries. The model for integrase family strand exchange illustrated above (Fig. 2c; 13) is consistent with the measured ΔLk values, and with the knotting/catenation topologies of the products, assuming that synapsis by 'random collision' accounts for the diversity in the products. In λ Int-catalysed *att*P versus *att*B recombination, some interdomainal nodes might be trapped by DNA wrapping in the 'intasome' (46).

3.3 The effects of supercoiling

A requirement for optimal recombination by enzymes of the resolvase family is that the substrate is negatively supercoiled. For the inversion systems, e.g. *gix*/Gin, the requirement seems to be absolute (9), whereas for Tn3 resolvase, reactions of non-supercoiled substrates can be observed, but are slower than those with supercoiled molecules (13, 47) (see Section 4.3). Most integrase family reactions are less dependent on supercoiling, although it is important for normal Int-catalysed integrative recombination (48), and it affects product topology in these systems (Section 4.4; 4.9).

In general, any process that alters the geometry of part of the DNA double helix (e.g. by bending, twisting, or melting it) is likely to be influenced by the supercoiling state of the molecule. The effects of supercoiling in biological systems have been reviewed recently (4). In site-specific recombination, every stage in the reaction is likely to be affected in some way by supercoiling. Because supercoiling increases the plectonemic wrapping of the DNA, 'random collisions' (Fig. 8b) tend to give more complex synapse topologies at higher levels of supercoiling (49). Formation of complex synaptic structures might be energetically favoured or disfavoured by changes in supercoiling (Section 4.2; 50) and likewise, the thermodynamics of strand exchange might be strongly affected (Section 4.2; e.g. ref. 10).

4. Models for topological selectivity

4.1 Introduction

Here we discuss some of the ideas that have been put forward to explain topological selectivity. An important issue is how selection of a single synapse topology can be achieved. Some points need to be emphasized at the outset. First, one must distinguish initial synapsis from productive synapsis. Initial synapsis may be defined as the first energetically significant contact between the two sites on a pathway that may lead to recombination. The topology of the initial synapse may or may not be preserved until strand exchange. The productive synapse is the structure which is present when catalysis of strand exchange takes place. It is the topology of the productive synapse (i.e. the topological relationship of the two crossover sites) that contributes to final product topology, and that must therefore be fixed for Tn3 resolution and Gin/*gix* inversion. Much of the speculation about selectivity mechanisms implies that the topology after initial synapsis is fixed, or that the final, productive synapse topology is related to initial synapse topology by a precise transformation. This may be true, but should not be assumed (see Section 4.2.).

It is very unlikely that the rates of the processes leading to initial synapsis are significant in controlling the overall rate of recombination in the usual substrates. The data of Shore *et al.* (51) suggest that four-nucleotide restriction fragment ends interact cohesively on average several times (at least) per second, when separated by 1–2 kbp on a linear DNA molecule. A similar rate is predicted by theoretical analysis (52). The effective local concentrations of DNA segments on supercoiled molecules are probably much higher than on linear ones (4). There seems to be no reason, then, why initial synapsis should not normally be fast. However, the optimal half-time for overall recombination is usually 1–5 min. This suggests either that there is a low equilibrium concentration of functional initial synapses (due to instability, or because many of them are incorrectly formed 'dead ends') or that there is one or more slow steps before or after initial synapsis. A slow late step *could* be the conversion of initial to productive synapse, or at strand exchange (53). The relative rates of different pathways for initial synapsis could nevertheless in principle be significant in determining partition between possible productive synapse topologies, and the mechanism of initial synapsis might be critical for structural changes required to form the productive synapse.

The 'default model' for initial synapsis is called random collision. We use this term to imply that there is not any ordered mechanism for bringing the sites together, i.e. that they meet by three-dimensional diffusion. This should not be taken to mean that all ways of bringing the sites together are equally likely. In a simple case, two sites on the same molecule might be more likely to collide than sites on different molecules, due to enhanced local concentration. In circular non-supercoiled molecules, collisions with simple initial synapse topologies would be expected to be most likely because of the energetics of introducing distortion of the path of the DNA axis. Likewise, supercoiled DNA molecules appear to have an equilibrium solution form that is plectonemic (Section 3.2). Random collisions might then be most likely to entrap either zero nodes, or negative interdomainal 'supercoil' nodes (Fig. 8b).

4.2 Models for *res*/resolvase topological selectivity

4.2.1 Tracking

A conceptually straightforward mechanism for ensuring a unique synapse topology is tracking. It was proposed that resolvase bound at *res* forms a protein–DNA complex that 'loops' onto adjacent DNA, and scans it for another *res* site by one-dimensional diffusion (Fig. 11a; 27). By this process, interdomainal nodes could be eliminated, and a unique synapse topology could be guaranteed. This model provides an explanation for selection against intermolecular reactions and reactions of inverted sites, and an observed selection for recombination between adjacent *res* sites in substrates with four directly repeated sites (54–56). The searching complex would be expected to contact an inverted *res* differently from a directly repeated *res*, would not contact sites on other molecules at all, and would contact adjacent sites first.

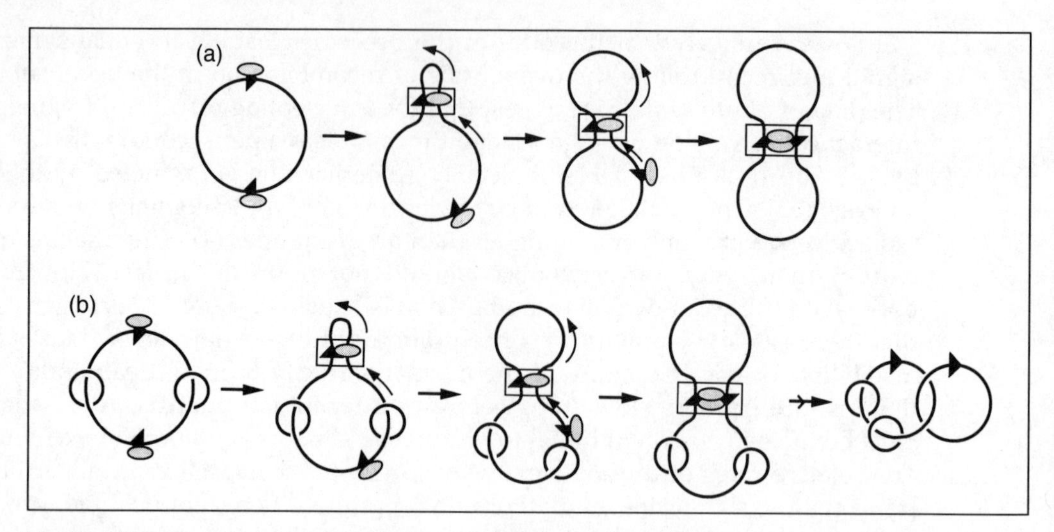

Fig. 11 (a) Tracking model for *res* synapsis. The shaded ovals represent resolvase bound at *res* (black triangles). (b) 'Reporter ring' test of the tracking model (54–56). The figure shows how the tracking would segregate the catenated DNA rings to one of the two circles of the product catenane. This was *not* observed.

However, tracking is not now considered adequate to explain resolvase's selectivity. The most elegant refutation of the simple tracking model was the 'reporter ring' experiment (54–56) (Fig. 11b). Tracking would segregate 'reporter rings' catenated to the substrate to only one of the two DNA domains in the synapse, and thence to only one of the two resolution product circles for any single substrate molecule. Random distribution of the circles was observed. Several other experiments (47, 57) are also inconsistent with tracking synapsis. All similar models where a continuous path is traced along the DNA between the sites are likewise unsatisfactory.

4.2.2 Slithering

On the demise of tracking for *res*/resolvase, two other models for the system's topological and resolution selectivity were proposed. One model (54, 55) postulated that an important type of diffusion in supercoiled DNA is 'slithering', a one-dimensional, conveyor-belt-like motion. At its simplest a plasmid would be limited to the behaviour cartooned in Fig. 12. It is proposed that all productive interactions of *res* sites are by slithering initial synapsis. By constraining the sites to meet in this way, a fixed initial synapse topology can be ensured. The original slithering model proposed an interaction of toroidally wrapped *res* sites to give the required ($S = -3{,}0$) productive synapse topology. It has now been shown that this type of synaptic geometry is incorrect; the (-3) topology involves plectonemic interwinding (13, 58; Fig. 10). None the less, the sites could still synapse by slithering. Slithering in branched plectonemic DNA is proposed to account for adjacent site selectivity in four-site substrates (see above), and for three-site synapsis as is needed for inversion reactions (Section 4.3). (Slithering in a linear plectoneme

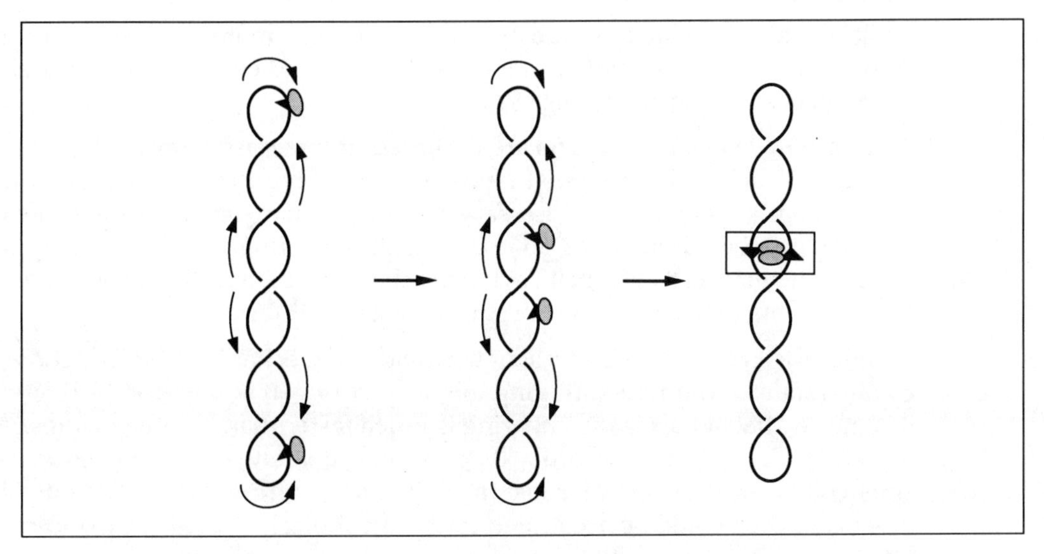

Fig. 12 Slithering model for *res* synpasis in supercoiled substrates (54, 55).

could not give adjacent site selectivity or three-site synapses.)

Like tracking, slithering accounts for selection against recombination between inverted or intermolecular sites by implying that encounters will be very infrequent, or in an unusable geometry.

At the moment there is no independent biochemical, structural, or theoretical evidence to support the idea that diffusion in supercoiled DNA is primarily by slithering. Although slithering may be a kinetically significant diffusion process, it is inadequate to account for resolvase's topological selectivity for the following reasons.

1. The slithering model is specific to supercoiled DNA. Resolvase can catalyse recombination between sites in non-supercoiled substrates, although the reactions are slower than in supercoiled molecules. Topological selectivity is evident in non-supercoiled substrates despite the absence of any possibility for restricting topology by slithering. For example, the main product from a direct repeat circular nicked substrate is the (−2) catenane, as it is in supercoiled substrates (Fig. 7b), and reactions of directly repeated sites are still strongly favoured over intermolecular or inverted sites in nicked molecules (47; M. R. Boocock, unpublished results). Therefore slithering is not required for recombination, nor is it sufficient to explain topological selectivity. [The reasons for the reduced reaction rates in the absence of supercoiling are not clear, but there are two obvious candidates for steps that might be supercoiling-dependent. One is the interwrapping of the *res* sites required to form a productive synapse (Section 4.2.3), which might be favoured by negative supercoiling. A second is strand exchange itself, which might normally be 'driven' by loss of supercoils (see, for example, ref. 13)].

2. Reactions of *res* sites in substrates with four or more *res* sites gave preferred products that were not in agreement with predictions of either a simple or 'branched' slithering model (59).
3. 'Random collision' products are observed in integrase family reactions (Section 3.2). So such collisions must happen reasonably often. If the slithering model is to be maintained, it is necessary to propose that somehow the protein–DNA complex formed at *res* (or *gix*) restricts its searching to DNA passing it on the plectonemic 'conveyor belt', whereas that formed at *att*P (for example) selects *against* targets that approach it in this way.

Slithering as proposed does not provide a mechanism for selection against recombination from non-slithering initial synapses. It is therefore a kinetic model: it suggests that synapsis by slithering is much faster than by other routes. A recent report by Parker and Halford (60) presented evidence that Tn*3* *res* synapsis in supercoiled substrates is very fast ($t_{1/2} < 10$ ms) and irreversible. The rate observed was faster than could be accounted for by theoretical models of three-dimensional diffusion in DNA, and slithering was invoked as a possible reduced dimensionality diffusion process that could account for the rate. However, a statistical mechanics study of supercoiled DNA (72) led to the conclusion that slithering is slow relative to other modes of diffusion.

4.2.3 Topological filter (two-step synapsis) (10, 47, 61)

The premise of this model is that catalysis of strand exchange by resolvase is only possible in a complex protein–DNA structure with fixed local geometry. It is hypothesized that the formation of this structure is dictated by DNA topology. Although initial synapses may be formed with many different topologies, only one initial topology allows the productive synapse to be formed correctly. Unsatisfactory initial synapses can dissociate and 'try again'. A similar model has been proposed for topological selectivity in Mu transposition (62).

Integral to this model, then, are proposals for the geometry and structure of the synaptic complex. The geometry of the DNA in the complex has been inferred from topological data and predictions of bending preferences within *res* from sequence analysis (10). It is proposed that the three negative interdomainal nodes required between the crossover sites are trapped by interwinding of the two *res* sites, mediated by resolvase intersubunit interactions.

A possible scheme for formation of the complex is as follows (Fig. 13a).

1. Resolvase subunits bind to the three subsites of *res*. The simple hypothesis illustrated in Fig. 13a, is that each subsite binds two subunits as a dimer, but it should be stressed that there are other possibilities. The subunits might interact further with each other to form a 'resolvosome' (9) but this is not required for the model.
2. Two *res* sites meet, and interact by antiparallel pairing of subsites II and III of each *res*. The paired sites then interwrap as depicted in Fig. 13a. Again it is

economical to speculate that the pairing and interwrapping are mediated by resolvase dimer–dimer interactions.

3. The two subsites I come together in a 'parallel' alignment by interactions of resolvase subunits bound at each, to form the productive synaptic complex that can catalyse strand exchange.

It is proposed that, in a circular substrate, only initial synapses with (S = 0,0) topology (i.e. where the two loops of DNA attached to the synapse are unlinked) can accommodate the interwrapping of the two *res* sites (steps 2 and 3 above) that transforms the initial synapse of the accessory subsites into the productive (S = −3,0) synapse of the crossover subsites. The term 'two-step synapsis' refers to these two distinct topological steps in the synapsis pathway (47, 61).

Consider first the proposed pathway for a productive synapsis (Fig. 13a). At the first interaction of resolvase subunits, the synapse topology is (S = 0,0). The DNA domains on either side of the point of contact are not interlinked. Now resolvase-mediated interactions between the accessory subsites II and III lead to interwinding of the two *res* sites. Finally the pair of crossover sites come together. It is only at this point that the synapse topology can be described as (−3,0), since there are three negative nodes trapped at the accessory sites if the point of synapsis is now regarded to be at the interacting parallel-aligned crossover sites.

Now consider an alternative scheme, where the first contact between the *res* sites has (S = −2,0) topology (Fig. 13b). The same series of events as in the first case now causes tangling of the DNA outside the synapse. The tangling itself, and the positive writhe that is introduced, are predicted to be very energetically costly. This tangling is expected to be more unfavourable in a supercoiled molecule than in a relaxed, nicked, or linear one (see below).

Similar tangling problems arise in any non-zero initial synapses (i.e. $S \neq 0,0$). In a substrate with inverted repeat *res* sites, it is not possible to align the pair of subsites II and III in antiparallel without introducing at least one interdomainal node. So again formation of the productive synapse would cause tangling. In intermolecular synapses of *res* sites on circular molecules, the sites can be aligned without interdomainal nodes, but interwinding introduces six nodes in total (Fig. 13c).

Another possible opportunity for selectivity is at strand exchange. Even if the final synaptic complex is formed from a non-zero initial synapse (e.g. as in Fig. 13c), strand exchange might no longer be coupled to a favourable free energy change due to loss of supercoiling.

An important distinction between the slithering and topological filter models is that in the slithering model selection is at initial synapsis, and is dependent on the mechanism of the interaction; in the topological filter model, selection is after initial synapsis, and is independent of the mechanism of the initial interaction. Boocock *et al.* (47, 61) favour random collision initial synapsis, followed by a topological filter, because there is then harmony with the results from the integrase family reactions, recombination selectivities in multisite substrates can be explained (59), and there is no need to advocate any 'special' diffusion properties of DNA.

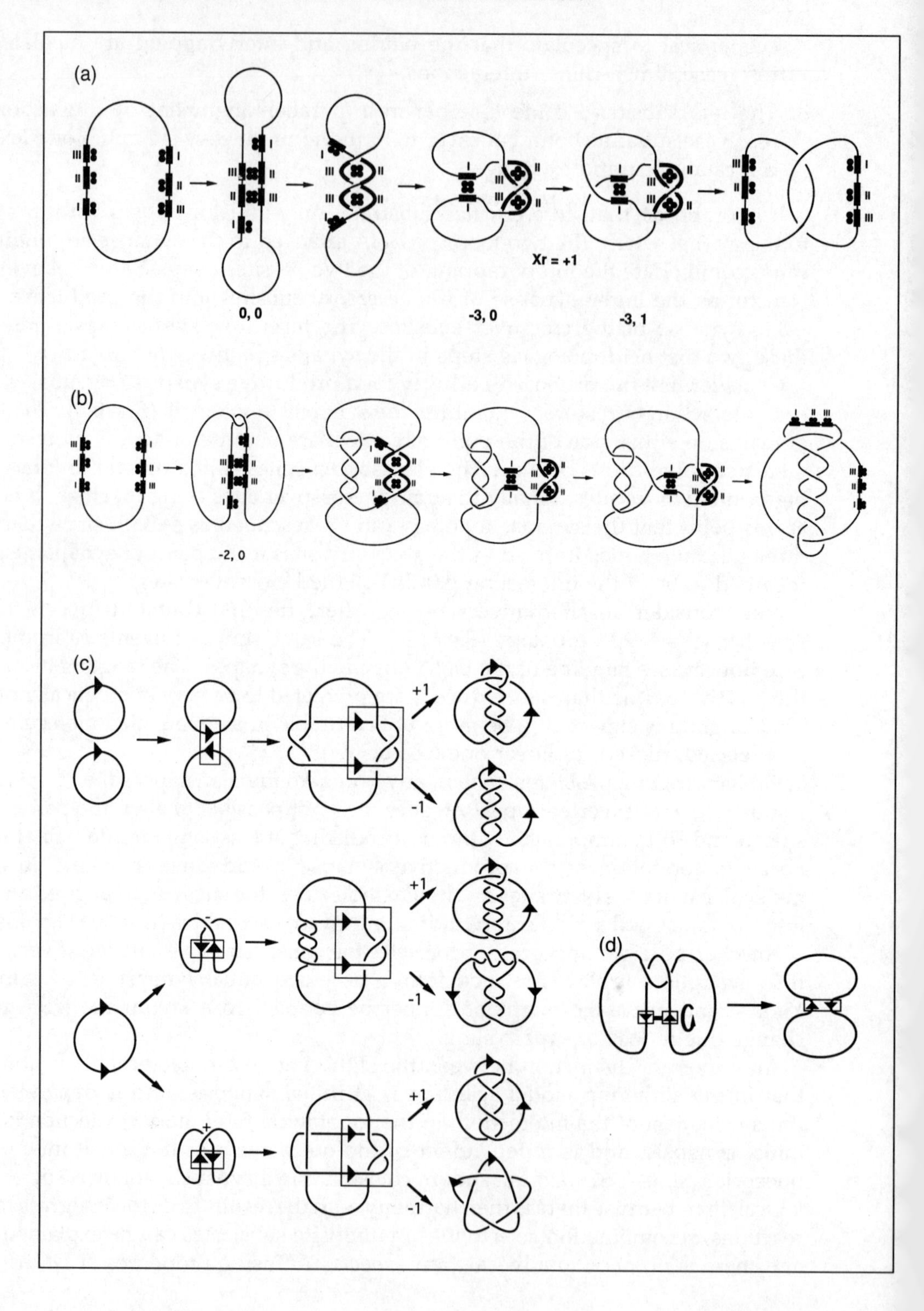
(a)
0, 0
-3, 0
Xr = +1
-3, 1
(b)
-2, 0
(c)
+1
-1
+1
-1
-
+
+1
-1
(d)

Fig. 13 Two-step synapsis (topological filter) model for *res*/resolvase topological selectivity. (a) Initial synapse with zero interdomainal nodes ($S = 0,0$). (b) Figure showing how initial synapsis trapping interdomainal nodes (in this case two negative ones) would lead to tangling of the DNA in the productive synapse, and complex catenation of the recombinant circles. The product shown is not observed. (c) Possible topologies of fusion (top) and inversion (bottom) reactions, via a 'productive synapse' of the proposed structure (inside the rectangular boxes). For each productive synapse, the outcomes from right-handed ($Xr = +1$) and left-handed ($Xr = -1$) strand exchange are shown (see Fig. 9). The products shown are not observed in reactions of supercoiled substrates. (d) How an initial synapse with non-zero topology (here $S = -2,0$) might be convertible into one with $S = 0,0$ if there is rotational freedom at the synapse.

Resolution can be quantitative. The topological filter model therefore requires an 'escape route' for abortive initial synapses with $S \neq 0,0$. The obvious possibility is that they simply dissociate rapidly and 'try again' by another random collision event. An alternative that must be considered is that inappropriate initial synapses can be converted into ($S = 0,0$), ones, without dissociation. This might be possible if there were rotational freedom at the point of initial synapsis, because most non-zero synapses are expected to entrap plectonemic supercoil nodes only (Fig. 13d). Experiments with multisite substrates tend to support rapid dissociation (59). Other evidence suggests that synapses quickly become very stable (60). It is possible that unstable interactions precede formation of the observed stable intermediate.

Another factor to be considered is that even with 'random collision' there might be a bias towards the intramolecular ($S = 0,0$) synapse. This is consistent with evidence from topoisomerase and integrase family recombinase reactions (e.g. ref. 49). A rationalization would be that non-zero synapses require an energetically unfavourable distortion of the axis of the DNA plectoneme. The bias should be especially marked in substrates with relatively closely spaced sites.

There is a further complication to the situation with inverted repeat sites. An isolated subsite I of *res* can recombine (albeit slowly) with a whole *res* site, giving resolution products even when the sites are in inverted repeat (22). Yet a pair of inverted full *res* sites do not recombine. The implication is that there is a further bar to inversion reactions of wild-type sites. It is possible that the sites form a relatively stable synapse that is, however, completely unproductive. For example, a parallel alignment of the two sites such that subsites I, II, and III of each *res* can pair, might do this. There is direct experimental evidence for synapsis of inverted *res* sites (60, 63; Mark Watson, unpublished results) and also evidence from multisite experiments that pairs of inverted *res* sites can have abnormally low reactivity with other sites (59).

The proposed topological barriers to synapsis of intermolecular or inverted sites could be reduced or removed by nicking or linearizing the substrates, and this has been shown to be the case (47; M. R. Boocock, unpublished results). Directly repeated sites still recombined much better than inverted repeat sites in nicked circles, but inversion reactions were now observable. Intermolecular recombination of nicked circular molecules was observable but very slow, whereas linear molecules were much better substrates for intermolecular reaction. The intermolecular reaction between a *res* on a linear molecule and one on a supercoiled circle can be quite efficient. Intramolecular reactions of *res* sites on linear molecules showed loss of selectivity for direct repeat sites, as predicted. The 'forbidden' reactions (intermolecular and inversion) of nicked circles should still produced tangling in the synapse, but it should not be so bad energetically because of the absence of supercoiling. All the topologically characterized nicked circle and nicked catenane reactions are consistent with a strong requirement for a synaptic complex of the proposed geometry (Fig. 13). For example, resolution of nicked circles gives the two-noded catenane, and inversion gives a five-noded knot, as predicted (47; M. R. Boocock, unpublished).

The barriers to recombination between inverted *res* sites, or sites on separate supercoiled circles, can also be reduced by multiple catenation or knotting of the substrates (58, 64). The results are again consistent with the proposed geometry of the synapse as in Fig. 13, but the specific substrate topologies used in these experiments allow formation of the synapse without further entanglement of the rest of the DNA.

A recent study of reactions between *res* sites on separate circles of nicked two-noded catenanes (73) strongly supported a topological filter mechanism for resolvase's selectivity.

4.3 Topological selectivity of DNA invertases

The normal reactions catalysed by Gin and Hin give exclusively unknotted inversion products, and ΔLk (for Gin) $= +4$, consistent with ($S = -2,0$) synapse topology and right-handed simple rotation strand exchange (Section 3.2). It would be nice to be able to propose a topological filter model that would also explain the unique topology of these systems, but there is no obvious simple extrapolation from *res*/resolvase. *Gix*, *hix*, and *cix* do not have accessory subsites. However, the requirement for an enhancer element *in cis* and for FIS protein suggests that together they might have an important role in selectivity (9, 32).

Reactions of substrates with directly repeated *gix* or *hix* sites only very low yields of recombinant resolution products, but result in extensive knotting on the plasmids (19, 21). The topologies of the knots are consistent with double rounds of 'simple rotation' strand exchange, in a synapse with ($S \times -2,0$) topology. The interpretation is that, as with *res*/resolvase, a synaptic with ($S = -2,0$) topology. The interpretation is that, as with *res*/resolvase, a synaptic structure with specific local geometry is required for catalysis of strand exchange, but that the left–right assymmetry of the inversion sites is not detected until after commitment to strand exchange. Therefore a ($-2,0$) synapse is formed in both direct and inverted repeat substrates. In direct repeat substrates, a single 180° simple rotation would then create mismatched base-pairs in the recombinants (Section 2.2), whereas the observed products of apparent 360° rotation are matched but non-recombinant.

A fascinating development has been the isolation of mutants of Cin and Gin that no longer require the enhancer and FIS for activity. The mutants were selected by screening for FIS- or enhancer-independent inversion, but they turn out to have lost other selectivities (65, 66). They no longer require supercoiling for recombination, and they recombine sites in direct or inverted repeat or on separate molecules. Productive synapse topology is also no longer fixed (74). Although requirement for FIS and the enhancer has been lost, they still modulate the reactions of the Gin mutants, tending to encourage 'wild-type' behaviour, i.e. intramolecular inversion with (−2,0) topology (65). Second mutations in these proteins can restore the dependence on enhancer/FIS (75).

Does the enhancer/FIS interact directly with the synapsed *gix* sites and recombinase, or does it act at a distance to promote certain types of synapsis of the *gix* sites? The balance of evidence suggests that there is direct interaction of FIS and the enhancer with the synapse. Crosslinking and electron microscopy with immuno-staining of Hin substrates revealed complexes containing the paired *hix* sites and enhancer together with Hin and FIS protein (67). The *hix* sites can also pair without the enhancer/FIX, suggesting that the enhancer/FIS might interact with a performed complex of Hin and the *hix* sites. However, the knotting reactions of some mismatched (directly repeated) *gix* and *hix* substrates are not compatible with catalysis within a fixed recombinase/*gix* (or *hix*)/enhancer + FIS complex. It is possible that after synapsis and commitment to strand exchange, the enhancer element is free to leave the complex (19, 21).

One hypothesis for the mechanisms of selection of (−2,0) synapse topology is that the three sites (2 × *gix* + enhancer) meet at a three-way junction is branched plectonemic DNA, by slithering (30, 68; Fig. 14).

A topological filter model might predict that productive synapsis requires pairing of two *gix* sites with initial (S × 0,0) topology as in the case of *res*/resolvase. The selection might similarly be based on the distortion of the DNA caused by interwrapping of the sites. The problem is how this synapse might be converted into one with productive (S = −2,0) topology. Simple inspection (Fig. 14b) indicates that the enhancer element or another segment of DNA would have to pass through the synapsed *gix* sites to achieve this, if Gin and *gix* maintain their initial mode of interaction. This problem can be avoided if one *gix* must first associate with the enhancer before the second *gix* arrives. The electron microscopy evidence, however, suggests that for *hix*/Hin the first association is between two *hix* sites (67).

There are two other possibilities if the enhancer is to interact with a performed (2 *gix*) + Gin complex. One is that the enhancer + FIS can induce a rearrangement that converts the initial synapse to (*S* = −2,0) *without* passing the enhancer through the complex. This could involve introducing a solenoidal supercoil at each *gix* site (Fig. 14b). Alternatively, the *gix*/Gin complex could have (−2,0) topology before interaction with the enhancer. This could be selected by solenoidal wrapping of the individual *gix* sites (as in Fig. 14b), or it could be that the enhancer only interacts productively with initial *gix* synapses of (−2,0) topology, i.e. the 'topological filter' could involve geometric requirements for the interaction of the enhancer/FIS and *gix*/Gin. Recent evidence from experiments with enhancer-independent Gin (69) suggests that the enhancer/FIS may catalyse a conformational

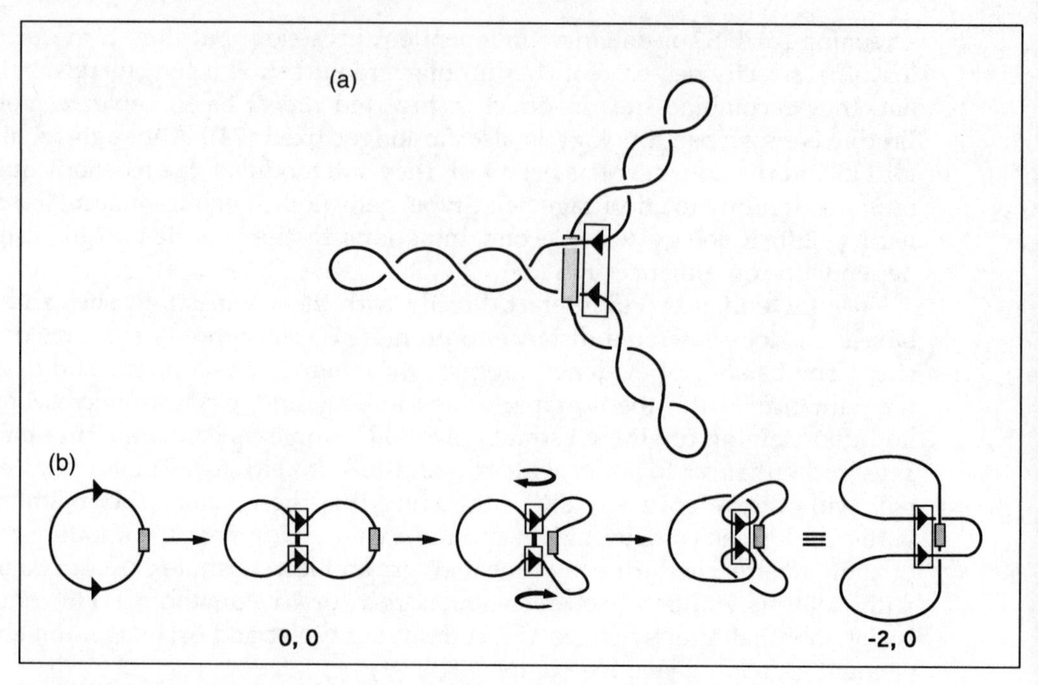

Fig. 14 (a) Three-site synapsis at a branch point in plectonemically supercoiled DNA, as proposed for Gin/*gix* inversion (30, 68). The *gix* sites are represented by black arrowheads, and the enhancer by a shaded rectangle. (b) A speculation about how the enhancer and FIS might induce a rearrangement of a Gin/*gix* synapse with $S = 0,0$ into one with $S = -2,0$.

change in synapsed *gix*/Gin complexes, causing unwinding of the *gix* DNA. It is proposed that in the natural reaction, enhancer/FIS acts to promote this change only in the (−2,0) synapse, thus activating Gin for strand exchange, whereas the mutant Gin can bring about the conformational change without enhancer/FIS, and therefore shows loss of topological selectivity.

It should be noted that there is a thermodynamic reason why inversion might be more efficient with non-zero synapsis. 'Simple rotation' strand exchange in a synapse with ($S = 0,0$) topology has no associated linkage change ($\Delta Lk = 0$) and so is not coupled to loss of supercoils. The observed ΔLk of +4 of the reaction [presumably with ($S = -2,0$) synapsis] might thermodynamically assist strand exchange.

4.4 Topological selectivity in integrase family reactions

Intramolecular Int-catalysed reactions may have a wide range of product topologies, which have been interpreted as being formed by random collision of sites (35, 41). In supercoiled molecules, the product knots or catenanes (for inversion or resolution reactions respectively) belong to the torus class, and are consistent with random entrapment of plectonemic interdomainal supercoil nodes in the synapse (46). Non-supercoiled molecules (nicked circles) give simpler products; in these substrates random collision would trap fewer nodes on average, and they might be (+) or (−) in sign. In λ Int *att*P versus *att*B inversion, there is selective formation of (+) trefoil knot rather than unknotted products from nicked circles (46, 70), suggesting that the DNA in the *att* sites may be wrapped in some way. The Int family recombinases FLP and Cre make unknotted circles as the simplest inversion product. In agreement with the correlation of topological selectivity with connection selectivity in the resolvase family, these integrase family reactions show little selectivity for particular connections of sites; resolution (excision), inversion, and fusion (integration) are all allowed in appropriate substrates (39, 40).

The topologically complex products of some integrase family reactions indicate that synapses with non-zero topology, formed by random collision, can lead to a productive synapse and strand exchange. This might suggest that less interwrapping of the two sites is needed, or that the recombinase is 'stickier', or less demanding of precise synaptic geometry, than resolvase family enzymes.

Integrase family reactions can give topologically restricted products (Section 3.2.4). This altered selectivity can be induced by changing the reaction conditions, mutations in the sites or recombinase, or, in the case of λ Int, by omitting Xis from the reaction. It is not at all clear how these changes affect the properties of the reactions in this way. Such easy switching between apparent 'random collision' and 'restricted' synapsis suggests that there might be nothing fundamentally different between the 'selective' and the 'promiscuous' recombination systems in their mechanisms of site pairing. Further study of systems whose topological selectivity can be modulated might be very helpful in distinguishing between the models for selectivity discussed above.

5. Conclusions

The selectivity of resolvase for recombination only of directly repeated sites on the same circular molecule, and for formation of a product of unique topology (a simple catenane) turn out to be aspects of the same phenomenon, i.e. selection of unique topologies for productive synapsis and strand exchange. The invertase enzymes' selectivities for inverted sites and unknotted circle recombination products are explicable in the same way. In contrast, λ Int and its relatives in many circumstances do not specify unique synapse topology, and consequently do not show selectivity for particular types of site connection or for a single product topology.

The basis of the selection by resolvases and invertases of unique productive synapse topologies remains controversial. Of the several models that have been proposed, we favour the 'topological filter' as most satisfactorily accommodating the available evidence. However, it is still not clear how this model might fully explain the results from the invertase systems. The issues might eventually be resolved by critical experiments to test hypotheses on the diffusion behaviour of DNA itself, as well as by direct *in vitro* studies on site-specific recombination systems.

References

1. Sadowski, P. D. (1993) Site-specific genetic recombination: hops, flips, and flops. *FASEB J.*, **7**, 760.
2. Gellert, M. and Nash, H. A. (1987) Communication between segments of DNA during site-specific recombination. *Nature*, **325**, 401.
3. Stark, W. M., Boocock, M. R., and Sherratt, D. J. (1992) Catalysis by site-specific recombinases. *Trends Genet.*, **8**, 432.
4. Kanaar, R. and Cozzarelli, N. R. (1992) Roles of supercoiled DNA structure in DNA transactions. *Curr. Opin. Struct. Biol.*, **2**, 369.
5. Mizuuchi, K. (1992) Transpositional recombination: mechanistic insights from studies of Mu and other elements. *Annu. Rev. Biochem.*, **61**, 1011.
6. Mack, A., Sauer, B., Abremski, K., and Hoess, R. (1992) Stoichiometry of the Cre recombinase bound to the *lox* recombining site. *Nucleic Acids Res.*, **20**, 4451.
7. Landy, A. (1989) Dynamic, structural, and regulatory aspects of λ site-specific recombination. *Annu. Rev. Biochem.*, **58**, 913.
8. Nash, H. A. (1990) Bending and supercoiling of DNA at the attachment site of bacteriophage λ. *Trends Biochem. Sci.*, **15**, 222.
9. Hatfull, G. F. and Grindley, N. D. F. (1988) Resolvases and DNA invertases: a family of enzymes active in site-specific recombination. In *Genetic Recombination*. R. Kucherlapati and G. R. Smith (eds). American Society for Microbiology, Washington, DC, p. 357.
10. Stark, W. M., Boocock, M. R., and Sherratt, D. J. (1989) Site-specific recombination by Tn*3* resolvase. *Trends Genet.*, **5**, 304.
11. Argos, P., Landy, A., Abremski, K., Egan, J. B., Haggard-Ljungquist, E., Hoess, R. H., Kahn, M. L., Kalionis, B., Narayama, S. V. L., Pierson, L. S., III, Sternberg, N., and Leong, J. M. (1986) The integrase family of site-specific recombinases: regional similarities and global diversity. *EMBO J.*, **5**, 433.
12. Abremski, K. E. and Hoess, R. H. (1992) Evidence for a second conserved arginine residue in the integrase family of recombination proteins. *Protein Engineering*, **3**, 87.
13. Stark, W. M., Sherratt, D. J., and Boocock, M. R. (1989) Site-specific recombination by Tn*3* resolvase: topological changes in the forward and reverse reactions. *Cell*, **58**, 779.
14. Weisberg, R. A., Enquist, L. W., Foeller, C., and Landy, A. (1983) Role for DNA homology in site-specific recombination. The isolation and characterization of a site affinity mutant of coliphage λ. *J. Mol. Biol.*, **170**, 319.
15. Nash, H. A. and Robertson, C. A. (1989) Heteroduplex substrates for bacteriophage lambda site-specific recombination: cleavage and strand transfer products. *EMBO J.*, **8**, 3523.
16. Kitts, P. A. and Nash, H. A. (1987) Homology-dependent interactions in phage λ site-specific recombination. *Nature*, **329**, 346.
17. Hoess, R. H., Wierzbicki, A., and Abremski, K. (1986) The role of the *lox*P spacer region in P1 site-specific recombination. *Nucleic Acids Res.*, **14**, 2287.
18. Senecoff, J. F. and Cox, M. M. (1986) Directionality in FLP protein-promoted site-specific recombination is mediated by DNA–DNA pairing. *J. Biol. Chem.*, **261**, 7380.
19. Kanaar, R., Shekhtman, E., Klippel, A., Dungan, J. M., Kahmann, R., and Cozzarelli, N. R. (1990) Processive recombination by the phage Mu Gin system: implications for the mechanism of DNA strand exchange, DNA site alignment, and enhancer action. *Cell*, **62**, 353.
20. Stark, W. M., Grindley, N. D. F., Hatfull, G. F., and Boocock, M. R. (1991) Resolvase-

catalysed reactions between *res* sites differing in the central dinucleotide of subsite I. *EMBO J.*, **10,** 3541.
21. Heichmann, K. A., Moskowitz, I. P. G., and Johnson, R. C. (1991) Configuration of DNA strands and mechanism of strand exchange in the Hin invertasome as revealed by analysis of recombinant knots. *Genes Dev.*, **5,** 1622.
22. Bednarz, A. L., Boocock, M. R., and Sherratt, D. J. (1990) Determinants of correct *res* site alignment in site-specific recombination by Tn3 resolvase. *Genes Dev.*, **4,** 2366.
23. Cozzarelli, N. R., Boles, T. C., and White, J. H. (1990) Primer on the topology and geometry of DNA supercoiling. In *DNA Topology and its Biological Effects*. N. R. Cozzarelli (ed.). Cold Spring Harbor Laboratory Press, Cold Spring Harbor, NY, p. 139.
24. Bellomy, G. R. and Record, M. T. (1990) Stable DNA loops *in vivo* and *in vitro*: roles in gene regulation at a distance and in biophysical characterization of DNA. *Prog. Nucleic Acids Res. Mol. Biol.*, **30,** 81.
25. Sherratt, D. J. (1989) Tn3 and related transposable elements: site-specific recombination and transposition. In *Mobile DNA*. D. E. Berg and M. M. Howe (eds). American Society for Microbiology, Washington, DC, p. 163.
26. Reed, R. R. (1981) Transposon-mediated site-specific recombination: a defined *in vitro* system. *Cell*, **25,** 713.
27. Krasnow, M. A. and Cozzarelli, N. R. (1983) Site-specific relaxation and recombination by the Tn3 resolvase: recognition of the DNA path between oriented *res* sites. *Cell*, **32,** 1313.
28. Wasserman, S. A. and Cozzarelli, N. R. (1985) Determination of the stereostructure of the product of Tn3 resolvase by a general method. *Proc. Natl Acad. Sci. USA*, **82,** 1079.
29. Kahmann, R., Mertens, G., Klippel, A., Brauer, B., Rudt, F., and Koch, C. (1987) The mechanism of G inversion. In *DNA Replication and Recombination*. T. J. Kelly and R. McMacken (eds). Alan R. Liss, New York, p. 681.
30. Kanaar, R., van de Putte, P., and Cozzarelli, N. R. (1988) Gin-mediated DNA inversion: product structure and the mechanism of strand exchange. *Proc. Natl Acad. Sci. USA*, **85,** 752.
31. Johnson, R. C., Bruist, M. B., Glaccum, M. B., and Simon, M. I. (1984) *In vitro* analysis of Hin-mediated site-specific recombination. *Cold Spring Harbor Symp. Quant. Biol.*, **49,** 751.
32. Johnson, R. C. (1991) Mechanism of site-specific DNA inversion in bacteria. *Curr. Opin. Genet. Dev.*, **1,** 404.
33. Adrian, M., Ten Heggeler-Bordier, B., Wahli, N., Stasiak, A. Z., Stasiak, A., and Dubochet, J. (1990) Direct visualization of supercoiled DNA molecules in solution. *EMBO J.*, **9,** 4551.
34. Torbet, J. and Di Capua, E. (1989) Supercoiled DNA is interwound in liquid crystalline solutions. *EMBO J.*, **8,** 4351.
35. Pollock, T. J. and Nash, H. A. (1983) Knotting of DNA caused by a genetic rearrangement. *J. Mol. Biol.*, **170,** 1.
36. Cozzarelli, N. R., Krasnow, M. A., Gerrard, S. P., and White, J. H. (1984) A topological treatment of recombination and topoisomerases. *Cold Spring Harbor Symp. Quant. Biol.*, **49,** 383.
37. Wasserman, S. A. and Cozzarelli, N. R. (1986) Biochemical topology: applications to DNA recombination and replication. *Science*, **232,** 951.
38. Wasserman, S. A., Dungan, J. M., and Cozzarelli, N. R. (1985) Discovery of a predicted DNA knot substantiates a model for site-specific recombination. *Science*, **229,** 171.

39. Hoess, R. H. and Abremski, K. (1985) Mechanism of strand cleavage and exchange in the Cre–*lox* site-specific recombination system. *J. Mol. Biol.*, **181,** 351.
40. Cox, M. M. (1988) FLP site-specific recombination system of *Saccharomyces cerevisiae*. In *Genetic Recombination*. R. Kucherlapati and G. R. Smith (eds). American Society for Microbiology, Washington, DC, p. 429.
41. Craig, N. L. and Nash, H. A. (1983) The mechanism of phage lambda site-specific recombination: collision versus sliding in *att* site juxtaposition. In *Mechanisms of DNA Replication and Recombination*. Vol. 10. N. R. Cozzarelli (ed.). Alan R. Liss, New York, p. 617.
42. Abremski, K., Frommer, B., Wierzbicki, A., and Hoess, R. H. (1988) Properties of a mutant Cre protein that alters the topological linkage of recombination products. *J. Mol. Biol.*, **202,** 59.
43. Nash, H. A. and Pollock, T. J. (1983) Site-specific recombination of bacteriophage lambda: the change in topological linking number associated with exchange of DNA strands. *J. Mol. Biol.*, **170,** 19.
44. Abremski, K., Frommer, B., and Hoess, R. H. (1986) Linking number changes in the DNA substrate during Cre-mediated *lox*P site-specific recombination. *J. Mol. Biol.*, **192,** 17.
45. Beatty, L. G., Babineau-Clary, D., Hogrefe, C., and Sadowski, P. D. (1986) FLP site-specific recombination of yeast 2 μm plasmid: topological features of the reaction. *J. Mol. Biol.*, **188,** 529.
46. Spengler, S. J., Stasiak, A., and Cozzarelli, N. R. (1985) The stereostructure of knots and catenanes produced by phage λ integrative recombination: implications for mechanism and DNA structure. *Cell*, **42,** 325.
47. Boocock, M. R., Brown, J. L., and Sherratt, D. J. (1987) Topological specificity in Tn3 resolvase catalysis. In *DNA Replication and Recombination*. T. J. Kelly and R. McMacken (eds). Alan R. Liss, New York, p. 703.
48. Richet, E., Abcarian, P., and Nash, H. A. (1986) The interaction of recombination proteins with supercoiled DNA: defining the role of supercoiling in lambda integrative recombination. *Cell*, **46,** 1011.
49. Abremski, K. and Hoess, R. H. (1985) Phage P1 Cre–*lox*P site-specific recombination: effects of DNA supercoiling on catenation and knotting of recombinant products. *J. Mol. Biol.*, **184,** 211.
50. Lim, H. M. and Simon, M. I. (1992) The role of negative supercoiling in Hin-mediated site-specific recombination. *J. Biol. Chem.*, **267,** 11176.
51. Shore, D., Langowski, J., and Baldwin, R. L. (1981) DNA flexibility studies by covalent closure of short fragments into circles. *Proc. Natl Acad. Sci. USA*, **78,** 4833.
52. Berg, O. G. (1984) Diffusion-controlled protein–DNA association: influence of segmental diffusion of the DNA. *Biopolymers*, **23,** 1869.
53. Meyer-Leon, L., Inman, R. B., and Cox, M. M. (1990) Characterisation of Holliday structures in FLP protein-promoted site-specific recombination. *Mol. Cell. Biol.*, **10,** 235.
54. Krasnow, M. A., Matzuk, M. M., Dungan, J. M., Benjamin, H. W., and Cozzarelli, N. R. (1983) Site-specific recombination by Tn3 resolvase: models for pairing of recombination sites. In *Mechanisms of DNA Replication and Recombination*, Vol. **10.** N. R. Cozzarelli (ed.). Alan R. Liss, New York, p. 637.
55. Benjamin, H. W. and Cozzarelli, N. R. (1986) DNA-directed synapsis in recombination: slithering and random collision of sites. *Proc. Robert A. Welch Found. Conf. Chem. Res.*, **29,** 107.

56. Benjamin, H. W., Matzuk, M. M., Krasnow, M. A., and Cozzarelli, N. R. (1985) Recombination site selection by Tn*3* resolvase: topological tests of a tracking mechanism. *Cell*, **40,** 147.
57. Saldanha, R., Flanagan, R., and Fennewald, M. (1987) Recombination by resolvase is inhibited by *lac* repressor simultaneously binding operators between *res* sites. *J. Mol. Biol.*, **196,** 505.
58. Benjamin, H. W. and Cozzarelli, N. R. (1990) Geometric arrangements of Tn*3* resolvase sites. *J. Biol. Chem.*, **265,** 6441.
59. Brown, J. L. (1986) Properties and action of Tn*3* resolvase. PhD thesis, University of Glasgow.
60. Parker, C. N. and Halford, S. E. (1991) Dynamics of long range interactions on DNA: the speed of synapsis during site-specific recombination by resolvase. *Cell*, **66,** 781.
61. Boocock, M. R., Brown, J. L., and Sherratt, D. J. (1986) Structural and catalytic properties of specific complexes between Tn*3* resolvase and the recombination site *res*. *Biochem. Soc. Trans.*, **14,** 214.
62. Craigie, R. and Mizuuchi, K. (1986) Role of DNA topology in Mu transposition: mechanism of sensing the relative orientation of two DNA segments. *Cell*, **45,** 793.
63. Benjamin, H. W. and Cozzarelli, N. R. (1988) Isolation and characterisation of the Tn*3* resolvase synaptic intermediate. *EMBO J.*, **7,** 1897.
64. Droge, P. and Cozzarelli, N. R. (1989) Recombination of knotted substrates by Tn*3* resolvase. *Proc. Natl Acad. Sci. USA*, **86,** 6062.
65. Klippel, A., Cloppenborg, K., and Kahmann, R. (1988) Isolation and characterization of unusual *gin* mutants. *EMBO J.*, **7,** 3983.
66. Haffter, P. and Bickle, T. A. (1988) Enhancer-independent mutants of the Cin recombinase have a relaxed topological specificity. *EMBO J.*, **7,** 3991.
67. Heichman, K. A. and Johnson, R. C. (1990) The Hin invertasome: protein-mediated joining of distant recombination sites at the enhancer. *Science*, **249,** 511.
68. Kanaar, R., van de Putte, P., and Cozzarelli, N. R. (1989) Gin-mediated recombination of catenated and knotted DNA substrates: implications for the mechanism of interaction between *cis*-acting sites. *Cell*, **58,** 147.
69. Klippel, A., Kanaar, R., Kahmann, R., and Cozzarelli, N. R. (1993) Analysis of strand exchange and DNA binding of enhancer-independent Gin recombinase mutants. *EMBO J.*, **12,** 1047.
70. Griffith, J. D. and Nash, H. A. (1985) Genetic rearrangement of DNA induces knots with a unique topology: implications for the mechanisms of synapsis and crossing-over. *Proc. Natl Acad. Sci. USA*, **82,** 3124.
71. Stark, W. M. and Boocock, M. R. (1994) The linkage change of a knotting reaction catalysed by Tn*3* resolvase. *J. Mol. Biol.*, **239,** 25.
72. Marko, J. F. and Siggia, E. D. (1994). Fluctuations and supercoiling of DNA. *Science*, **265,** 506.
73. Stark, W. M., Parker, C. N., Halford, S. E., and Boocock, M. R. (1994) Stereoselectivity of DNA catenane fusion by resolvase. *Nature*, **368,** 76.
74. Crisona, N. J., Kanaar, R., Gonzalez, T. N., Zechiedrich, E. L., Klippel, A., and Cozzarelli, N. R. (1994). Processive recombination by wild-type Gin and an enhancer-independent mutant. *J. Mol. Biol.*, **243,** 437.
75. Merker, P., Muskhelishvili, G., Deufel, A., Rusch, K., and Kahmann, R. (1993) Role of Gin and FIS in site-specific recombination. *Cold Spring Harbor Symp. Quant. Biol.*, **58,** 505.

6 | Transposable elements as selfish DNA

J. F. Y. BROOKFIELD

1. Introduction

Biologists seek functional explanations for the attributes of organisms. Such explanations show the advantage that the attribute gives to the organism. Any attribute, including components of the genome, can potentially be explained in this way. The belief that such functional explanations are appropriate rests on the tenet of the Neo-Darwinian evolutionary theory that states that the only systematic mechanism by which genes or other DNA sequences can spread through populations is through differences in the fitnesses of their carriers. Transposable elements, however, increase in number either directly or indirectly during transposition, and thus could potentially spread without increasing the fitness of their hosts. This non-Darwinian behaviour illegitimizes functional explanations and requires us to seek causal explanations for the presence of transposable elements in genomes, incorporating both their own molecular properties and their effects on their hosts. One class of explanations are those in which the effects of transposable elements on their hosts are negative. In other words, they are 'selfish DNAs' (1, 2). In this chapter, I will discuss some of the ideas and experiments indicating the selfishness or otherwise of transposable genetic elements. For details of the structures of various classes of transposable elements and the mechanisms of their transposition, the reader should consult other chapters in this volume. I will concentrate mainly on the elements in sexually reproducing diploid eukaryotes, for which the data sets of the distributions and frequencies of elements are often more complete, and for which the Neo-Darwinian population genetics synthesis provides a framework for considerations of transposable element evolution. I will strive, however, to highlight the ways in which the evolutionary process affecting prokaryotic transposable elements shows important similarities to and differences from the diploid paradigm.

2. Categorization of the mechanisms for the spread of transposable elements

Three classes of mechanisms can be proposed to account for the spread of transposable elements through host populations.

1. Transposable elements may convey a direct advantage on their carriers. In other words, the individuals in the population bearing transposable elements may have a higher Darwinian fitness than individuals lacking them. For example, bacterial transposons carrying antibiotic-resistance genes increase the fitness of their host cells in conditions where the antibiotic has been applied (3).
2. They give an advantage to the hosts by promoting genetic variability. This genetic variability may then be used by the host in its adaptation to a changing environment.
3. They replicate more quickly than other components of the genome and thereby increase their representation in the population. If their effects on the host are negative, then they are selfish DNAs. Here a distinction is drawn between selfishness and over-replication *per se*, since elements that are advantageous to their hosts may also over-replicate. It may be, however, that it is their advantageousness that plays the major role in causing their spread.

2.1 Overview of these mechanisms

These three mechanisms are not mutually exclusive, and may all operate simultaneously. The last seems to be the most important, for the following reasons. The first mechanism, while important in bacteria, is virtually unknown in eukaryotes. The second mechanism is highly problematical. Firstly, it should be noted that whatever the mechanism is by which transposable elements spread, they will promote genetic diversity in doing so, and thus the finding of such induced variability is not, in itself, evidence that it is causal in their spread. More importantly, this mechanism is not a true alternative to the other two. The generation of mutation by a transposable element will only bring about the spread of the element if the process results in individuals bearing the element being fitter than those that lack it. In other words, the spread still has to occur through the first mechanism. I will consider below conditions in which the induction of mutation by transposable elements would or would not have this consequence.

2.2 Advantageous transposable elements

Only a small minority of transposable elements, and all from bacteria, convey a selective advantage on their hosts. Some *Escherichia coli* transposons, such as Tn3, which carries an ampicillin-resistance gene, replicate in transposition (4). It is, nevertheless, clear that the spread of this element through a population of bacteria under conditions in which ampicillin is applied would occur whether transposition was or was not replicative. Thus one can view this situation as being one in which selection at the level of the host controls the spread of the element. There are, indeed, some examples in which insertion sequences, such as IS*50*, can improve the growth rate of their host cells, in the absence of transposition and through a mechanism which is still mysterious (5). It is hard to imagine mechanisms of this kind generating the situation seen in mammals, for example, in which transposable element sequence families may be represented by tens of thousands of individual copies in the genome. If it is possible to select strongly for individual copies of elements, it is not clear what would be the mechanism for increasing copy number when it is already very high.

2.3 Transposable elements as generators of diversity

2.3.1 Experimental support

Transposable elements generate a significant proportion of spontaneous mutations of major effect in a number of important genetic organisms such as *Saccharomyces cerevisiae* (e.g. 6, 7) and *Drosophila melanogaster* (e.g. 8–11). Thus they might also be expected to generate a high proportion of the mutations occurring in wild populations. There has been considerable debate as to whether the mutation rate in wild populations is optimal or minimal (12), but, if it is optimal, and a proportion of it is the result of the activities of transposable elements, then this implies an advantage conveyed by transposable elements to their hosts. However, the data set of mutations induced by transposable elements may be biased in that mutations studied by geneticists are chosen for their major phenotypic effects and a large proportion are null mutations. Such mutations are not the kind normally used in adaptive evolution. Genetic lineages do not evolve by eliminating more and more genetic functions by null mutations as the environment demands. Rather, gene functions are remarkably stable in evolution and are altered by subtle changes in expression and activity. Transposable genetic element insertions may, however, also have subtle effects. Transposable elements may insert some way from the coding sequences, and thus produce small phenotypic effects, by having minor effects on the transcription of nearby genes. For example, mobilization of the *D. melanogaster P* element by hybrid dysgenic crosses generates variation in quantitative characters which can subsequently be demonstrated by directional selection (13–15). One can thus imagine that at least a small proportion of transposable element insertions might convey a selective advantage on a host adapting to a new environment. This is what was found in a series of experiments in which *E. coli* and *S. cerevisiae* strains bearing transposable elements increased their fitness in chemostats more rapidly than strains which lack these elements (16–18). It must be said that similar chemostat experiments to those performed on *E. coli* were also found to favour single-copy alleles that induced increases in the mutation rate. Selection appears to favour any source of mutation as the population undergoes an evolutionary adaptation to the new and unnatural environment of the chemostat (19, 20).

2.3.2 Contrary evidence from *Drosophila*

In *D. melanogaster*, surveys using *in situ* hybridization to polytene chromosomes with labelled transposable element probes have shown that all transposable element sites are occupied at low frequencies in wild populations (reviewed in 21, 22). If transposable elements were making insertion mutations which were subsequently utilized in adaptive evolution, these insertions would become fixed in populations. The finding that all sites of transposable element insertions have low frequencies in populations thus implies that insertion mutations have not been used by *D. melanogaster* in its adaptive evolution. In other species, such as the mammals, individual transposable elements sites are usually fixed in populations (23, 24). However, since fixed sites could easily be generated by genetic drift, such sites are not, in this situation, evidence for adaptive substitutions, but rather for the smaller effective population size of mammals than of insects.

2.3.3 The generation of 'footprints' in the genome

One potential way around the problem of why transposable element insertions are apparently not utilized in adaptive evolution is suggested, in particular, by a series of mutations generated in plants (25, 26), *Drosophila* (27), and *Caenorhabditis elegans* (28) by elements which transpose via DNA intermediates. In these systems, the process of insertion and subsequent imprecise excision has been found to generate series of mutant alleles which differ from the pre-existing wild-type alleles by the addition or deletion of small numbers of bases. These changes, described as the 'footprints' of the transposable elements, are subtle at the molecular level, and could have effects on gene expression of the kind occasionally utilized in adaptive evolution. Clearly, mutations of this kind in *Drosophila* would not be detected by *in situ* hybridization. Such mutant alleles arising from imprecise excision of transposable elements may have a major role in evolution. However, the existence of this process does not explain the existence of transposable elements. The host individuals with the advantageous mutation no longer have the transposable element at the locus. Thus the spread of the mutant allele in the population will not spread the transposable element that initially generated it. This conclusion applies to a sexual population in which variable loci and transposable elements at other sites will be in linkage equilibrium. There will be situations, however, of positive linkage disequilibrium between the element (now located at another site) and the mutation, and in such cases the element may spread by a form of 'hitch-hiking'. For example, the reduced rate of genetic exchange in bacteria may promote the spread of element-bearing strains through insertion mutagenesis. (17).

2.4 Transposable elements as 'selfish' DNAs

2.4.1 Is transposition replicative?

The ability of a transposable element family to spread through populations of hosts relies, in this model, upon the element's ability to replicate more quickly than the host DNA around it. For many transposable elements, the process of transposition is clearly replicative. The retrotransposons and retroposons, which transpose via RNA intermediates, fall into this category. For other elements, the transposition process is conservative at the molecular level. Examples include the bacterial transposon Tn*10* (29), the *Zea mays Activator* (*Ac*) element (30, 31), and the *D. melanogaster P* element (32). However, a process may be conservative at the molecular level and yet have the over-replication of the element as its consequence. Thus, the *Ac* element transposes from replicated DNA, and thus, by frequently inserting in DNA which has yet to be replicated, has the opportunity of two rounds of DNA replication in a single S phase. The *E. coli* transposon Tn*10* also transposes out of hemimethylated and thus recently replicated DNA. The *P* element leaves behind a double-strand break in the donor chromosome, which is repaired by gene conversion from the homologue, thereby again generating a net replication of the element (32). This mechanism strongly recalls the molecularly conservative yet evolutionarily replicative behaviour of the *E. coli* transposon Tn*5* (33). These mechanisms are diagrammatically illustrated in Fig. 1. The ubiquity of such effectively replicative transposition processes indicates that transposable elements will be capable of spreading, even in the absence of any selective advantage to their carriers.

2.4.2 The genome as an ecological community

While the concept of parasitic or selfish DNA can potentially explain how transposable elements can spread through host populations, many important questions remain, some of which I will discuss below. Examples include what prevents transposable elements from causing the extinction of their hosts, and what determines how much DNA falls into the 'selfish' class? The selfish DNA hypothesis predicts that a genome includes a set of competing evolutionary lineages, those of the host and the various families of selfish DNA. The competition process between these lineages is analogous to that studied by the interface between evolution and community ecology, which seeks to understand how the interactions between coexisting species in natural communities are affected by evolutionary changes within each species. It is depressing to note that, despite much effort, the problem of understanding this interface has remained unsolved. Similar difficulties have been encountered in the other analogous situation of the interactions between viruses and their hosts. Our understanding of immunology and analogous host defence systems, and of the viral evolutionary process, is still not sufficient to make accurate predictions of the co-evolution of the host–virus interaction. There seems no compelling reason to imagine that the co-evolutionary process of transposable elements and their hosts will be any less complex.

2.4.3 Ignorant DNA

It is also hard to classify the possibility that extra DNA may exist which is weakly harmful or has no effect, but which has been generated by a biased amplification process which, in other circumstances, is advantageous to the host, such as in the

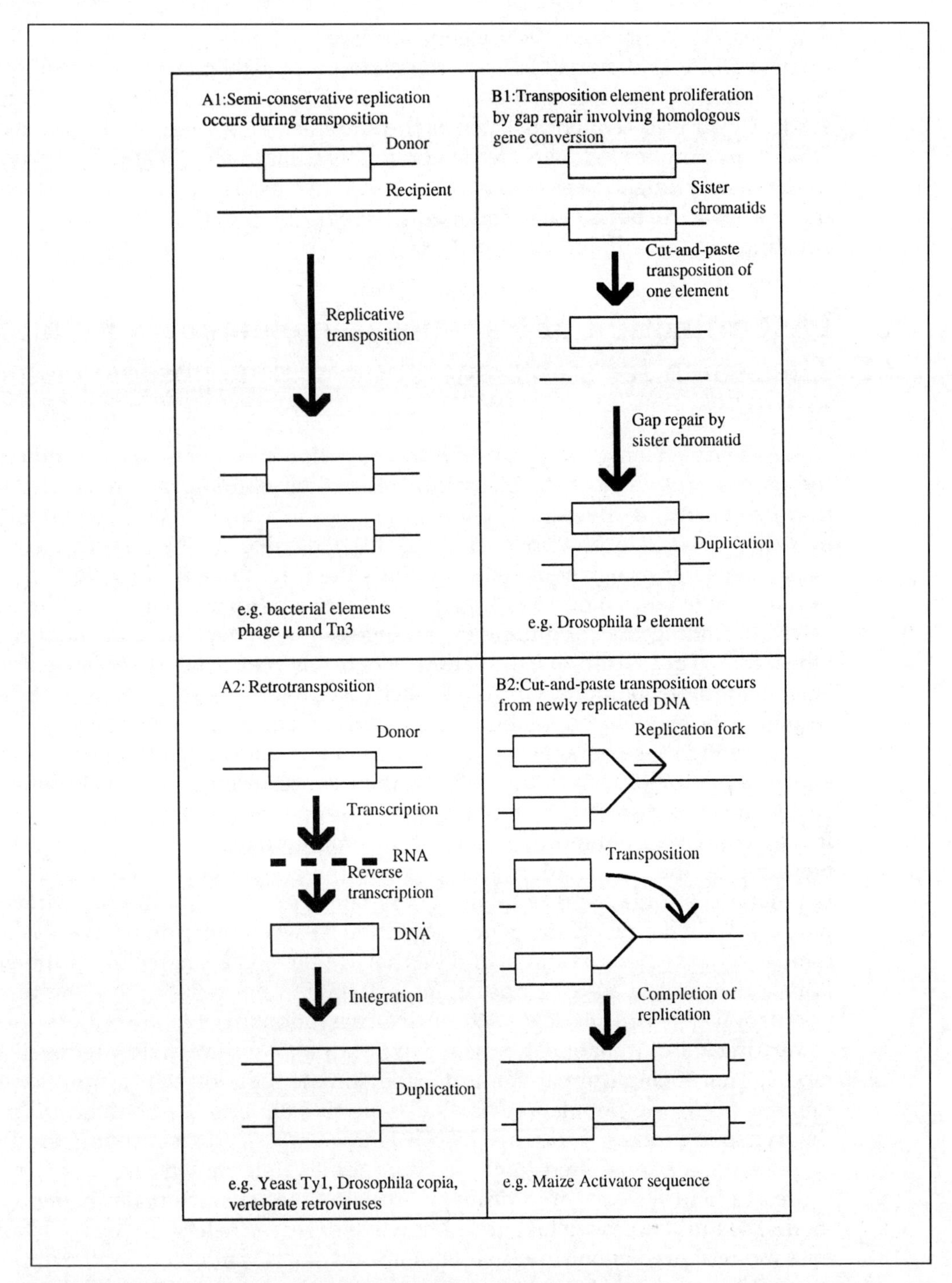

Fig. 1 Four ways in which transposition can lead to an increase in transposable element copy number. For further details of the replicative transposition mechanism see Chapter 2.

amplification of ribosomal RNA genes or histone genes (34). Amplified but functionless DNA, which may include the majority of satellite sequences, may be the coincidental by-product of the existence of enzymes catalysing amplifications of tandemly arrayed sequences, such as the amplifications found to restore ribosomal DNA copy number following reduction in *D. melanogaster* (35) and *S. cerevisiae* (36). In the case of functionless DNAs of this kind, in contrast to transposable elements, there appear to be no sequence requirements in order for a DNA to act as a substrate for the amplification process.

3. Determination of transposable element copy number

3.1 The search for stable equilibria in transposable element abundances

If transposable elements replicate in transposition, then, if they have no effect on the fitness of their host, they must increase their abundance when rare. Even if they have harmful effects on the host, there is still a large area of parameter space, in which their rate of transposition is high relative to the selective coefficient against hosts bearing them, which allows them to spread. However, theoretical population genetics models of transposable element invasion must not limit themselves to finding the conditions for an increase in transposable element numbers when rare. The nature of population genetics is that it takes the measured parameters of the evolutionary process, such as mutation rates, selective coefficients, recombination rates, population sizes, and so on, and makes dynamic models incorporating these parameters. These dynamic models must then be subjected to experimental or observational tests. If they concern long-term evolutionary processes they can be tested by looking at related species or subspecies of organisms. If they concern evolution over very short time periods, then they can potentially be tested by the examination of changes in the genetic composition of laboratory populations of organisms with short generation times. More often, however, and particularly in models of the genetic variation between individuals, there is no time dimension in the observations and all that we have is a 'snapshot' of the population, a description of the population at a single time point. Thus the testing of population genetics models with such observational data can only be done by comparing the population 'snapshot' with an equilibrium state predicted by the model. That such a comparison is itself legitimate relies on the assumption that in the real world the population has had time to evolve to the equilibrium point of the dynamic process through which it is evolving. This assumption is a big one because populations often have their properties determined largely by historical processes, and we cannot in principle explain these characteristics using dynamic models without access to historical knowledge. Nevertheless, that population genetics models' predictions are their equilibrium states reminds us of the importance of examining models for such equilibria rather than merely for their dynamics. In the context of transposable element models, the major theoretical problem is not whether or not harmful transposable elements can invade host populations by over-replication but what stops the invasion process, and thereby prevents the mean transposable element number from increasing without limit. The simplest dynamic models of selfish elements assume that each transposable element has a constant probability of replicative transposition, and that each extra element lowers host fitness by the same proportional degree. Perhaps counter-intuitively, in sexual species these models yield, at most, two stable equilibrium points. These are, respectively, zero transposable elements per individual, and maximum host fitness, and an infinite number of transposable elements, and a host fitness of zero (37). Obviously, neither state corresponds to the observed situation of transposable elements co-existing with their hosts. One major theoretical problem has been, therefore, the finding of a mechanism that can stabilize copy number at some intermediate value.

3.2 Self-regulation of transposable elements

Since we cannot see a population in which transposable elements have reduced host fitness to zero, it follows that this biased ascertainment implies that any transposable element we see will have some system for the stabilization of copy number. That this is so, however, is not equivalent to a demonstration that such regulation will evolve. Many transposable elements, particularly in prokaryotes, are known to self-regulate. There has been considerable theoretical effort applied to the problem of self-regulation of transposition in sexual species, without any powerful and general mechanism which would cause the evolution of self-regulation to occur having been found (38). One can easily imagine a transposable element which lowers its transposition rate when the copy number in the host is high, and one can see that such behaviour would stabilize the co-existence of the element and the host. It is, however, hard to see how such an element could spread through the transposable element population when it is the only element which displays such restraint. The fundamental problem is that a transposable element with a lowered capacity for transposition, be this through changes in *trans-* or *cis*-acting functions, will transpose less quickly than typical elements of its family. This effect will thus slow its replication. In a sexual species, such a transposable element will not, however, be found in individuals with reduced transposable element copy numbers. This is because, in each generation, recombination and fertilization will randomize the genetic background of the element, which will thus be found in individuals with copy numbers typical of the population as a whole. Thus, since it undergoes less replicative transposition without any compensating advantage, an element with reduced activity, even if this is only manifest when copy number is high, will not spread and will be eliminated from the transposable element population. This argument applies less strongly in an asexual species. Indeed, it is very hard to imagine selfish transposable elements spreading at all in a purely asexual species. A host lineage lacking all such elements will be fitter and will inexorably replace all the lineages bearing elements, however rapid the elements' rates of transposition within those lineages may be (39).

The argument that self-regulation will not evolve also assumes that it is not transposition itself which generates the harmful effects of transposable elements. The models assume that the harmful effects of transposable elements come from the harmful effects of increased element copy number, or from mutations which do not act as dominant lethals. If the harmful effects of transposition come immediately, as the direct consequence of transposition itself, then elements with reduced transposition rates may be able to spread to replace more active elements (38, 40).

3.3 Stable equilibria between transposition and selection

The model for selection outlined above, which fails to stabilize copy number, is of an equal proportional reduction in fitness with each extra element present in the genome. Apart from simplicity, there seems no biological reason to justify the adoption of such a model for fitness. Theory predicts that a stable equilibrium between replicative transposition and selection can be reached if the effects of increased copy number on fitness increase more than multiplicatively. One suggestion (21, 22, 41) is that a major selective force that acts against transposable elements in eukaryotes is the generation of aneuploid gametes through ectopic recombinations between members of the same transposable element family at different chromosomal locations. This suggestion has two advantages. Firstly, since the ectopic recombination events will involve two transposable elements, their frequency would be expected to increase with the square of the copy number, and thus the relationship between copy number and fitness would be of the stabilizing kind. Secondly, the theory can be subjected to an experimental test. We have detailed knowledge of the distribution of recombination rates in the *Drosophila* chromosomes, with the centromeric and telomeric regions displaying a reduced rate of recombination. One can infer that the rates of ectopic recombination involving these chromosomal regions could be correspondingly reduced. The model would therefore predict that elements located in these regions would be subject to weaker selection and that their equilibrium numbers would be higher. Such an effect is indeed found for the majority of elements tested. A number of elements tend to show highly significant excesses in abundance in the basal euchromatin of the X and other chromosomes (41, 42).

4. Variation in transposable element positions in wild populations

4.1 Methods for detecting variation in transposable element positions

The majority of the evidence concerning the variation between host individuals in the positions of their transposable elements has come from four groups, *Drosophila* (21, 22), *S. cerevisiae* (43–45), *E. coli* (46, 47), and mammals (23, 24). Data concerning transposable element positions come from three sources. Firstly, transposable elements can be detected by probing Southern blots of restriction-digested genomic DNA with unique sequence probes. Individual transposable elements are manifested as insertion–deletion polymorphisms or, if they are fixed, by insertion–deletion variation between related species, by the sequencing of the cloned region, or by hybridization of the clones to probes of the transposable element family. This technique is the only one used in mammals, and has been used for various genomic regions in *Drosophila* (reviewed in ref. 22). Secondly, in *Drosophila*, the technique of *in situ* hybridization to polytene salivary gland chromosomes has been greatly utilized. This technique allows the simultaneous visualization of all transposable element copies at sites in euchromatin. Thirdly, in organisms in which copy number is low, one can obtain 'fingerprints' by probing, with the transposable element probe, genomic DNA cut with an enzyme which does not cut within the transposable element (48). Each element copy will generate a single band of hybridization, and, since fragment length will be the result of the distance to flanking restriction sites, elements of the same family at different sites will produce fragments of different lengths, which can thus be counted. Movements of the element can be observed by the loss and gain of new bands of hybridization.

4.2 The levels of variability detected

The most interesting and general observation from the *Drosophila* data is that all transposable element sites are occupied at low frequencies in wild populations. In other words, there is enormous variation between hosts in the positions of their transposable elements. This result is inconsistent with their having advantageous effects which are position-dependent but consistent with their being selfish DNAs, which are subject either to a deletion process or weak selection at individual sites. A distribution which has been called the frequency spectrum of transposable element sites can be easily produced from the *in situ* hybridization data. This is the observed relationship between the number of sites in the chromosome that have a frequency of x in the sample of chromosomes, and x. This frequency spectrum can be compared with that expected according to the theory of neutral mutation (49). Early models differed subtly in the precise form of the neutral theory that was adopted. One postulated that sites were always generated by unique insertions (50) while another assumed that repeated transpositions into a given site were possible (37). Both models give good fits to the observed frequency spectra, implying that there is no significant evidence for multiple insertions at the same site. The fit between theory and observation allows the numerical parameters of the models to be estimated from the data. In particular, it is possible to estimate the product of the rate of transposition per generation and the effective population size of the host population. Estimates for various elements (reviewed in ref. 22) range from three upwards, but many elements have yielded infinite estimates of this product (a result which follows from all sites being represented uniquely in the sample). Even the finite values calculated are associated with very large standard errors. Early data (51) were also consistent with a model in which there were no sites shared by different genomes, and in which apparently shared sites were the result of the inaccuracy of the *in situ* technique (52). Later evidence has proved to be inconsistent with this model. Since these models are based on the assumption of a stable equilibrium between replicative transposition and a contrary force limiting transposition numbers, the finding of high variation implies that either deletion or selection acts on transposable element sites with a force per generation that is stronger than the reciprocal of the host's effective population size. This does not necessarily indicate absolutely high rates of transposition or of selection. The effective population size of *D. melanogaster* in the eastern USA, from which the populations used were sampled, has been estimated at 2×10^4 by lethal allelism studies (53).

The mammalian data differ greatly from those of *Drosophila*. Mammalian interspersed repetitive DNAs, all of which are transposable, have been divided into SINEs (short interspersed elements) and LINEs (long interspersed elements) on the basis of length and on the basis of the presence in LINE sequences of genes encoding transposition proteins, particularly reverse transcriptase. The vast majority of the sites of LINEs and SINEs are fixed in the host population. The element sites can be seen to be lost or gained when related species such as man, chimpanzee, and gorilla (54) or different species of *Mus* (55) are compared.

In *E. coli*, studies of the positions of a collection of insertion sequences (IS*1*, IS2, IS*3*, IS*4*, IS*5*, and IS*30*) revealed very great differences between strains from a collection of 71 reference strains (47). Copy numbers for the individual IS families ranged from 0 to 20 in the strains, and the variance in numbers between the strains was very much more than the approximately Poisson variance observed between individuals in *Drosophila* populations. This was as expected, since the low rate of recombination in *E. coli* populations (56) would be expected to generate positive associations between elements of the same family. Indeed, successful modelling of the IS element distributions using branching process theory produced predictions which fitted the data well (57). As in the *Drosophila* case, these models assume selfishness and a stable equilibrium being formed between selection and replicative transposition. The argument of Hickey (39) that selfish transposons cannot spread in clonal organisms can be countered by the ability of zero copy number lineages to acquire insertion sequences by infection on plasmids. Sequencing of insertion sequences IS*1*, IS*3*, and IS*30* has revealed very high sequence similarity of each of these elements throughout the range of *E. coli* isolates, indicating the rapid movement of insertion sequences among clonal lineages (58).

The *Drosophila* data sets can be examined in other ways to look for evidence of selection. As mentioned above, the generation of ectopic recombinations between different transposable element families could plausibly act as a mechanism for stabilizing a transposition–selection balance. Another form of selection which will limit transposable element numbers is the direct selection arising from the harmful effects of mutations generated by insertion. If weakly harmful alleles generated by transposable element insertions are recessive in their effects, they should be eliminated more rapidly on the X chromosomes, where they will be exposed to hemizygous selection every third generation. However, the data sets show no consistent under-representation of transposable elements on the X chromosome, which argues against their creating harmful recessive mutations with appreciable frequencies (59).

One interesting idea as to why it is possible for mammals to acquire thousands of copies of the same transposable element family, while *Drosophila* never has more than 100 or so copies of any family, is that it seems probable that ectopic pairing between transposable elements would be reduced by homozygosity at the individual sites. Thus, once site frequencies have become high by drift in a small population, an element will normally have an 'available' copy on the homologue to pair with at meiosis, and ectopic pairing could be reduced (41). This would reduce selection against the elements and thus provides a possible causal link between the high site frequencies and the high genomic copy numbers found for mammalian elements.

5. Evolutionary changes in transposable element sequences

5.1 Selection for transposability will not operate at any given site

If transposable elements are selfish DNAs, the selective constraint which acts upon them will act to maintain the sequences required for transposition. Any given transposable element insertion will initially be found in one individual in the population, and the insertion will spread through the population, if at all, only by genetic drift, possibly opposed by the forces of selection and excision. The selective constraint for transposability will not operate at any given site. Thus at a particular site, the sequence of a retrotransposon will not be subject to any selective constraint at all. Its evolution will be entirely neutral. This is because any mutations which destroy its function as a transposable element will affect only its expected contribution to the population of daughter elements being generated at new sites. In the case of an element transposing via an excision mechanism, mutations which prevent this excision process will give the mutated element an advantage at that site. As a result of this lack of constraint one must, therefore, expect many pseudogenes for transposable elements to exist. The proportion of family elements which are pseudogenes will depend on the frequency with which elements are selected for transposability compared with the mutation rate. The *Drosophila* retrotransposons are subject to a selection process which results in all insertion sites having low frequencies. This process will affect non-transposable pseudogenes equally, and individual sites of insertion will exist for short time periods. Thus all elements will be at sites recently generated by transposition and pseudogenes will be rare. Alternatively, elements with sites which are usually fixed in populations, such as the mammalian LINEs, will have occupied their current chromosomal sites for considerable periods, and will usually be pseudogenes, incapable of further transposition, a prediction confirmed by the sequences of mammalian retrotransposons studied (60).

5.2 The fate of non-autonomous elements

The above discussion concerns the fate of mutations that eliminate the *cis*-acting capacity of elements to transpose. One important aspect of the evolution of a transposable element family is whether deleted elements, possessing only the sequences required *in cis*, and not genes for *trans*-acting transposase proteins, can spread through the family. One important determinant of this is the extent to which non-autonomous transposable elements can be complemented *in trans* by other elements elsewhere in the genome. In many elements they obviously can be, and indeed deleted *D. melanogaster P* elements transpose at higher frequencies than do intact ones (61). Alternatively, if complementation is not absolute and transposase concentration is limiting for transposition rate, an intact copy will transpose more quickly than a deleted one. The interaction between transposition and selection may have a major effect on the success of deleted element copies. In a situation in which autonomous elements exist in a transposition–selection equilibrium, and in which fitness is a consequence of a total transposition rate, models predict that non-autonomous elements are generally able to invade a population and replace the autonomous elements (40), a prediction that is broadly consistent with the available data for the *D. melanogaster P* element (28, 62)

5.3 Recombination between transposable element copies

Transposable elements generally show very low degrees of specificity in their insertion sites. This means that the number of potential target sites in the chromosomes is very high and elements rarely insert repeatedly into the same site. Thus, if the same chromosomal site is occupied in two homologous chromosomes the two element copies will normally be descended from the same original insertion into the chromosome. Thus allelic recombination between transposable elements at meiosis will normally be between copies which have a recent common ancestor and thus are very similar in sequence. The possibility of recombination between more diverged members of a transposable element family could occur only by gene conversion (63) or, in the case of the retrotransposons, by the possible recombination within virus-like particles of two RNAs derived from different source copies (a situation observed experimentally with retroviruses (64), but expected to occur only at a low frequency under natural conditions). It seems probable that the effective levels of recombination between transposable element copies are very low.

5.4 Mutation rates and Muller's ratchet

Retrotransposons which lack the ability to make the proteins required for transposition are generally incapable of transposition, despite the presence of intact elements belonging to the same family in the same cell. This must indicate a packaging of elements such that RNA genomes can only be packaged with their own translation products. It is not obvious why such retrotransposons have not evolved to use transposition proteins encoded by other elements, and thereby be capable of transposing as non-autonomous elements, in the way that *P* elements and certain maize elements have (26). Retrotransposons and retroviruses will have very high mutation rates during the processes of transcription and reverse transposition (65, 66). Mutations generated during transposition, along with the unselected mutations generated in DNA replication during the many chromosomal generations between transpositions, will give a high level of mutation between successive transposition events. The length of retrotransposons in base-pairs may not be much less than the reciprocal of the mutation rate per base between transpositions. Non-recombining DNAs with mutation rates which are not orders of magnitude less than one will be subject to a process called Muller's ratchet (67). This is the process whereby, in a finite population, the random loss of all wild-type elements may occur. This means that all elements will have at least one harmful mutation, and further mutations may create a situation in which all elements bearing only single mutations may again become rare, and be randomly lost. This process may cause the gradual degeneration of a population of sequences, and this may limit the maximum length of retrotransposons and, indeed, retroviruses. Paradoxically, if mutations occurring in a sequence are only weakly deleterious, Muller's ratchet may operate in conditions in which it would not if all mutations were strongly deleterious. This is because weakly harmful mutations would have higher frequencies in a mutation–selection balance, and therefore a higher probability of chance fixation. A retrotransposon which always used its own transposase would leave daughter elements which were transpositionally competent whereas one which could be complemented, albeit inefficiently, *in trans* might not. It is tempting to speculate that the inability of retrotransposons to be complemented *in trans* is a mechanism to avoid Muller's ratchet. A subpopulation of retrotransposons which could be weakly complemented might suffer only weak selection against mutations destroying their own transposition genes, which might therefore be lost by Muller's ratchet. Retrotransposons which absolutely required their own transposition proteins would purge their population of mutations at each cycle of transposition and Muller's ratchet could be avoided.

6. The phylogenies of transposable elements

Any collection of homologous DNA sequences, such as members of a transposable element sequence family, if they show no recombination, must be connected by a phylogeny, which leads back to a single most recent common ancestor. The examination of extant sequences can be used to make inferences about the structure of this phylogeny, which can, in turn, be used to draw inferences about the evolutionary process by which the phylogeny was produced. Phylogenies of transposable element copies estimated from DNA sequences are now becoming available for a number of elements from diverse species (e.g. 68, 69).

6.1 Transposable element phylogeny illustrated by the mammalian SINE sequence Alu

If a number of copies of the same transposable element family are sequenced it is possible to produce an estimated phylogeny assuming parsimony. Such phylogenies can convey considerable information. As an illustration, around half a million copies of the Alu SINE sequence are found in the human genome, and similar numbers in our closest relatives (chimpanzee and gorilla). It is a retroposon of around 300 bp, which encodes no *trans*-acting products; it is merely a matter of semantics whether or not this sequence can be considered a transposable element. It is certainly mobile. Its origin was as a pseudogene of the 7SL signal recognition particle (70), which then underwent divergence followed by a fusion of two diverged elements (71). This dimeric element then increased in abundance and underwent further diversification. A large number of copies of this sequence derived from man, chimpanzee, and gorilla have now been sequenced, and estimates of their phylogeny have been produced.

6.1.1 Simple neutral models for homogenization of sequences do not fit the data

Assuming that all transposable element copies are functionally equivalent (i.e. their probabilities of transposition are equal), it is possible to produce a stochastic model predicting how long ago the transposable element copies had a common ancestor, and therefore their expected neutral DNA divergence (72–74). Such models, when applied to the Alu sequence, give unrealistically long times to common ancestors (74). This implies that the Alu copies are too similar in sequence to each other for the homogenization process simply to be random transposition and genetic drift of functionally equivalent sequences.

6.1.2 Possible alterations to the model

1. The Alu sequences may have undergone rapid expansion, and thus have not yet diverged from an ancestral sequence. This situation, which would be consistent with high copy number yet close similarity in sequence, has been treated theoretically (75). The theory for coalescent processes in an exponentially increasing population (76) is of relevance here and predicts that the sequences would approximately show a 'star' phylogeny, with all pairs compared having common ancestors at about the same time.

2. There may be frequent expansions, in the Alu sequence pool, of variants with enhanced transposition rates. While there might have been a high copy number for a long time, there could be a continuous turnover as improved Alu sequences arise and spread through the Alu sequence pool. This theory predicts that more recent, improved, subfamilies, could, while being rare in the Alu population as a whole, generate the majority of new Alu insertions.

3. All Alu transpositions could occur from a single or a small number of source copies. This suggestion has been investigated mathematically, and gives results which are consistent with typical Alu divergence patterns (77).

6.1.3 The sequence data reveal subfamily structure which supports the latter two models

The data set of human, chimpanzee, and gorilla Alu sequences has grown so large that the estimated phylogenies of the sequences can directly show the pattern of evolution of the family (24, 78–85). These phylogenies can be coupled with information on the timing of Alu insertions derived from comparisons between chimpanzee, human, and gorilla gene regions, in which Alu sequences can be seen to have inserted subsequent to the trichotomy separating the species. Furthermore, some Alu insertions can be inferred to be recent because they are polymorphic, or because they have created a visible mutation [such as a recent human neurofibromatosis Type 1 mutation (86)].

These data show that the phylogeny of human Alu sequences reveals a subfamily structure, rather than the 'star' phylogeny expected if all elements had recently expanded from a single source. However, it is still not clear whether the subfamily structure is the result of there being a small number of source genes, or of the occasional appearance and spread of new and improved types. The data reveal that most of the recent human Alu insertions have been of the subfamily PV, also known as HS-1, which differs from an earlier 'precise' subfamily by five tightly linked diagnostic mutations. This result can either be interpreted as resulting from mutations in a single source gene (78, 79, 80, 85) or from the spread a new improved sequence transposing at a high frequency from multiple source sites (82–84). It must be remembered that, since Alu sequences will decay by mutation at their chromosomal locations, the proportion of the total family which are active transposers may be very small. A new subfamily could form a large proportion of active copies, and thus give rise to a large proportion of new insertions, without forming more than a few percent of total Alu copies.

6.2 6.2 Phylogenies of the larger LINE elements

LINEs are longer (3–9 kb) elements which are found at high copy numbers in the chromosomes of mammals and other eukaryotes. They encode reverse transcriptases and evidently transpose via an RNA intermediate, yet lack the long terminal repeats of retrovirus proviruses and of more typical retrotransposons. Sequence information on LINE sequences from a number of mammals is now available (59, 80, 87–91). It has been possible to construct an estimated phylogeny of the L1 LINE sequence from three species of *Mus*, based on the base substitutions seen in a 320 base sequence of 32 sampled element copies (80, 88, 89). L1 sequences are usually pseudogenes, either because of 5′ deletions or base substitutions in the open reading frames. Even considering just the non-truncated copies, these authors found that all the bifurcations in the phylogenies could be accounted for if there were only one to three elements in the genome of each mouse species that were capable of acting as donor copies in transposition or gene conversion. They called these sequences 'molecular drivers'. However, since the number of sequences examined per species in the phylogeny is only 10, the theoretical maximum number of such 'driver' sequences required is only five. Even if all copies of the L1 sequence were fully functional it is highly probable that they would produce a phylogeny that would require three or fewer drivers. The data are probably consistent with there being up to a thousand active L1 sequences.

If repeated sequences of any type are to evolve and yet remain similar in sequence within a species, there must be concerted evolution, in which there are substitutions shared by all the element copies of a given family in a given host species. The cause of this is the shared descent of copies which inevitably results from the combined processes of transposition and genetic drift. Since the evolution of repeated gene families without the loss of sequence similarity between repeats logically implies concerted evolution, the invention of a terminology in which the observation of concerted evolution is transmogrified into its own cause ('molecular drive') (92) seems unnecessary.

6.3 Cases exist in which transposable element and host phylogenies differ

In the context of concerted evolution, there are interesting cases in which the phylogeny of transposable elements does not agree with that of their hosts. The *D. melanogaster P* element, which has spread through wild populations in this century (93), differs by a single base substitution from a sequence from the distantly related *D. willistoni*, which must be the original host (94). The mechanism of horizontal transfer between species is still uncertain, yet recent evidence implicates a parasitic mite (95). Similarly, phylogenies produced from the amino acid sequences of reverse transcriptase genes link transposable elements and retroviruses, but do not follow host phylogeny (96–98).

7. Speculations

7.1 Interactions between families

There appears to be a remarkably high degree of independence between transposable element families co-existing in the same host. This is perhaps surprising, since families will have had the opportunity to evolve in the cellular environments characterized by the products of other element families. In *D. melanogaster*, for example, early results suggesting mobilization of retrotransposons by P–M hybrid dysgenic crosses (99–101) have not been repeated in subsequent experiments (102). The implication is probably that certain interstrain crosses may weakly mobilize some retrotransposons, and that these crosses may, coincidentally, be P–M dysgenic. It seems likely that investigation of the relationships between co-existing families will be fruitful. It is, however, hard to find experimental tests for such interactions.

It is perhaps surprising that the diversity of co-existing element families, particularly in *Drosophila*, has turned out to be so high, with more than 20 families stably present in the same genome. One might have initially expected that one family would be a more efficient transposable element than the others and thus would have replaced them. The hypothesized selection arising from ectopic recombination mentioned above may hold the key to this question (41). Its consequence would be that members of a rare family of transposable elements would be subject to less selection than members of a common family, since they would encounter fewer elements with which they could ectopically pair. This frequency-dependent effect may maintain a rich community of diverse sequences in the same genome.

7.2 Transposable element origins

The similarities between the reverse transcriptase genes of the various families of retrotransposons and retroviruses from diverse eukaryotic hosts imply that the elements share common ancestry. Indeed, the data strongly suggest that the common ancestor of reverse transcriptase-encoding elements existed much more recently than did the common ancestor of the hosts. Furthermore, weak but significant homologies exist between the integrase domain of the pol protein from retroviruses and retrotransposons and the ORF B of bacterial insertion sequences (103, 104). This prompts speculation as to what the eukaryotic world was like prior to the existence of the common ancestor of these elements. Is the current collection of retro-elements merely the latest wave of such elements which were preceded by elements which were broadly similar in their mode of transposition, and their distribution and abundance? Alternatively, does the phylogeny of these elements back to their common ancestor reflect a spread of a new form of parasitic DNA sequence, and that, prior to this spread, no similar elements existed in eukaryotic chromosomes? If the latter is correct, then were there earlier forms of parasitic DNAs which were subsequently replaced by the modern types? There may be no way of answering these questions unless and until technology for studying the DNA from organisms that lived hundreds of millions of years ago becomes available.

7.3 The role of the host

In this discussion of parasitic transposable elements I have neglected discussion of the host's co-evolutionary process with the elements, and how evolution in the host has been able to eliminate transposable elements or limit the damage that they cause. The hosts have been seen as passive receptacles within which the transposable elements play out their own evolutionary process. This is probably far from accurate. Largely, this omission is the result of the ascertainment problem mentioned earlier. A host which had prevailed in its struggle against a particular family of parasitic transposable elements would long ago have eliminated them from its chromosomes and there would now be no interaction to study. However, another possibility is that the enzymic apparatus for DNA rearrangements, introduced into cells by parasitic transposable elements, could ultimately become used by hosts to carry out controlled and adaptive DNA rearrangements.

References

1. Doolittle, W. F. and Sapienza, C. (1980) Selfish genes, the phenotype paradigm and genome evolution. *Nature*, **284**, 601.
2. Orgel, L. G. and Crick, F. H. C. (1980) Selfish DNA: the ultimate parasite. *Nature*, **284**, 604.
3. Kleckner, N. (1981) Transposable elements in prokaryotes. *Annu. Rev. Genet.*, **15**, 341.
4. Arthur, A. and Sherratt, D. (1979) Dissection of the transposition process: a transposon-encoded site-specific recombination system. *Mol. Gen. Genet.*, **175**, 267.
5. Hartl, D. L., Dykhuizen, D. E., Miller, R., Green, L., and DeFramond, J. (1983) Transposable element IS50 improves growth rate of *E. coli* cells without transposition. *Cell*, **35**, 503.
6. Rose, M. and Winston, F. (1984) Identification of a Ty insertion within the coding region of the *S. cerevisiae URA3* gene. *Mol. Gen. Genet.*, **193**, 557.
7. Challeff, D. T. and Fink, G. R. (1980) Genetic events associated with an insertion mutation in yeast. *Cell*, **21**, 227.
8. Modolell, J., Bender, W., and Meselson, M. (1983) *Drosophila melanogaster* mutations suppressible by the *suppressor of hairy-wing* are insertions of a 7.3 kb mobile element. *Proc. Natl Acad. Sci. USA*, **80**, 1678.
9. Bender, W., Akam, M., Korch, F., Beachy, P. A., Peifer, M., Spierer, P., Lewis, E. B., and Hogness, D. (1983) Molecular genetics of the *bithorax* complex of *Drosophila melanogaster*. *Science*, **221**, 23.
10. Parkhurst, S. M. and Corces, V. G. (1986) Interactions among the *gypsy* transposable element and *yellow* and *Suppressor-of-hairy-wing* loci in *Drosophila melanogaster*. *Mol. Cell. Biol.*, **6**, 46.
11. Zachar, Z. and Bingham, P. M. (1982) Regulation of *white* locus expression: the structure of mutant alleles at the *white* locus of *Drosophila melanogaster*. *Cell*, **30**, 529.
12. Leigh, E. G., Jr (1973) The evolution of mutation rates. *Genetics*, **73** (Suppl.), 1.
13. Mackay, T. F. C. (1987) Transposable element-induced polygenic mutations in *Drosophila melanogaster*. *Genet. Res.*, **49**, 225.
14. Mackay, T. F. C., Lyman, R. F., and Jackson, M. S. (1992) Effects of P element insertions on quantitative traits in *Drosophila melanogaster*. *Genetics*, **130**, 315.
15. Torkamanzehi, A., Moran, C., and Nicolas, F. W. (1992) P element transposition contributes substantial new variation for a quantitative trait in *Drosophila melanogaster*. *Genetics*, **131**, 73.

16. Chao, L., Vargas, B., Spear, B. B., and Cox, E. C. (1983) Transposable elements as mutator genes in evolution. *Nature*, **303,** 633.
17. Chao, L. and McBroom, S. M. (1985) Evolution of transposable elements: an IS10 insertion increases fitness in *Escherichia coli*. *Mol. Biol. Evol.*, **2,** 359.
18. Wilke, C. M. and Adams, J. (1992) Fitness effects of Ty transposition in *Saccharomyces cerevisiae*. *Genetics*, **131,** 31.
19. Chao, L. and Cox, E. C. (1983) Competition between high and low mutating strains of *Escherichia coli*. *Evolution*, **37,** 125.
20. Gibson, T. C., Scheppe, M. L., and Cox, E. C. (1970) Fitness of an *Escherichia coli* mutator gene. *Science*, **169,** 686.
21. Charlesworth, B. and Langley, C. H. (1989) The population genetics of *Drosophila* transposable elements. *Annu. Rev. Genet.*, **23,** 251.
22. Charlesworth, B. and Langley, C. H. (1991) Population genetics of transposable elements in *Drosophila*. In *Evolution at the Molecular Level*. R. K. Selander, A. G. Clark, and T. S. Whittam (eds). Sinauer, Sunderland, Mass. p. 150.
23. Deininger, P. L. (1989) SINEs: short interspersed repeated DNA elements in higher eukaryotes. In *Mobile DNA*. D. E. Berg and M. M. Howe (eds). American Society for Microbiology, Washington, DC, p. 619.
24. Hutchinson, C. A., III, Hardies, S. C., Loeb, D. D., Shehee, W. R., and Edgell, M. H. (1989) LINEs and related retroposons: long interspersed repeated sequences in the eukaryotic genome. In *Mobile DNA*. D. E. Berg and M. M. Howe (eds). American Society for Microbiology, Washington, DC, p. 593.
25. Coen, E. S., Robbins, T. P., Almeida, J., Judson, A., and Carpenter, R. (1989) Consequences and mechanisms of transposition in *Antirrhinum majus*. In *Mobile DNA*. D. E. Berg and M. M. Howe (eds). American Society for Microbiology, Washington, DC, p. 413.
26. Federoff, N. (1989) Maize transposable elements. In *Mobile DNA*. D. E. Berg and M. M. Howe (eds). American Society for Microbiology, Washington, DC, p. 375.
27. Engels, W. R. (1989) P elements in *Drosophila melanogaster*. In *Mobile DNA*. D. E. Berg and M. M. Howe (eds). American Society for Microbiology, Washington, DC, p. 437.
28. Kiff, J. E., Moerman, D. G., Schriefer, L. A., and Waterston, R. H. (1988) Transposon-induced deletions in *unc-22* of *Caenorhabditis elegans* associated with almost normal gene activity. *Nature*, **331,** 631.
29. Bender, J. and Kleckner, N. (1986) Genetic evidence that Tn*10* transposes by a non-replicative mechanism. *Cell*, **45,** 801.
30. Gierl, A., Saedler, H., and Peterson, P. A. (1989) Maize transposable elements. *Annu. Rev. Genet.*, **23,** 71.
31. Greenblatt, I. M. and Brink, R. A. (1963) Transposition of Modulator in maize into divided and undivided chromosome segments. *Nature*, **197,** 412.
32. Gloor, G. B., Nassif, N. A., Johnson-Schlitz, D. M., Preston, C. R., and Engels, W. R. (1991) Targeted gap replacement in *Drosophila* via P element-induced gap repair. *Science*, **253,** 1110.
33. Berg, D. E., Berg, C. M., and Sasakawa, C. (1984) The bacterial transposon Tn*5*: evolutionary influences. *Mol. Biol. Evol.*, **1,** 411.
34. Dover, G. (1982) Ignorant DNA. *Nature*, **285,** 618.
35. de Cicco, D. V. and Glover, D. M. (1983) Amplification of rDNA and type I sequences in *Drosophila* males deficient in rRNA. *Cell*, **32,** 1217.
36. Kalback, D. B. and Halvorson, H. O. (1977) Magnification of genes coding for ribosomal RNA in *Saccharomyces cerevisiae*. *Proc. Natl Acad. Sci. USA*, **74,** 1177.

37. Charlesworth, B. and Charlesworth, D. (1983) The population dynamics of transposable elements. *Genet. Res.*, **42,** 1.
38. Charlesworth, B. and Langley, C. H. (1986) The evolution of self-regulated transposition of transposable elements. *Genetics*, **112,** 359.
39. Hickey, D. A. (1982) Selfish DNA: a sexually transmitted nuclear parasite. *Genetics*, **101,** 519.
40. Brookfield, J. F. Y. (1991) Models of repression of transposition in P–M hybrid dysgenesis by P cytotype and by zygotically-encoded repressor proteins. *Genetics*, **128,** 471.
41. Langley, C. H., Montgomery, E. A., Hudson, R., Kaplan, N., and Charlesworth, B. (1988) On the role of unequal exchange in the containment of transposable element copy number. *Genet. Res.*, **52,** 223.
42. Charlesworth, B. and Lapid, A. (1989) A study of ten families of transposable elements on X chromosomes from a population of *Drosophila melanogaster*. *Genet. Res.*, **54,** 113.
43. Cameron, J. R., Loh, E. Y., and Davis, R. W. (1979) Evidence for transposition of dispersed repetitive DNA families in yeast. *Cell*, **16,** 739.
44. Boeke, J. D. (1989) Transposable elements in *Saccharomyces cerevisiae*. In *Mobile DNA*. D. E. Berg and M. M. Howe (eds). American Society for Microbiology, Washington, DC, p. 335.
45. Boeke, J. D., Eichinger, D., Castrillon, D., and Fink, G. R. (1988) The *Saccharomyces cerevisiae* genome contains functional and non-functional copies of transposon Ty1. *Mol. Cell. Biol.*, **8,** 1432.
46. Ajioka, J. W. and Hartl, D. L. (1989) Population dynamics of transposable elements. In *Mobile DNA*. D. E. Berg and M. M. Howe (eds). American Society for Microbiology, Washington, DC, p. 939.
47. Sawyer, S., Dykhuizen, D., Dubose, R., Green, T., Mutangadura-Mhlanga, T., and Hartl, D. L. (1987) Distribution and abundance of insertion sequences among natural isolates of *Escherichia coli*. *Genetics*, **115,** 51.
48. Lawrence, J. G., Dykhuizen, D., Dubose, R. F., and Hartl, D. L. (1989) Phylogenetic analysis using insertion sequence fingerprinting in *Escherichia coli*. *Mol. Biol. Evol.*, **6,** 1.
49. Kimura, M. and Crow, J. F. (1964) The number of alleles that can be maintained in a finite population. *Genetics*, **49,** 725.
50. Langley, C. H., Brookfield, J. F. Y., and Kaplan, N. L. (1983) Transposable elements in Mendelian populations. I. A theory. *Genetics*, **104,** 457.
51. Montgomery, E. A. and Langley, C. H. (1983) Transposable elements in Mendelian populations: II Distribution of three copia-like elements in a natural population of *Drosophila melanogaster*. *Genetics*, **104,** 473.
52. Kaplan, N. L. and Brookfield, J. F. Y. (1983) Transposable elements in Mendelian populations. III Statistical results. *Genetics*, **104,** 485.
53. Mukai, T. and Yamaguchi, O. (1974) The genetic structure of natural populations of *Drosophila melanogaster*. XI Genetic variability in a local population. *Genetics*, **76,** 339.
54. Hwu, H. R., Roberts, J. W., Davidson, E. H., and Britten, R. J. (1986) Insertion and/or deletion of many repeated sequences in human and higher ape evolution. *Proc. Natl Acad. Sci. USA*, **83,** 3875.
55. Casavant, N. C., Hardies, S. C., Funk, F. D., Comer, M. B., Edgell, M. H., and Hutchinson, C. A., III (1988) Extensive movement of LINES one sequences in β-globin loci of *Mus caroli* and *Mus domesticus*. *Mol. Cell. Biol.*, **8,** 4669.
56. Hartl, D. L. and Dykhuizen, D. E. (1983) The population structure of *Escherichia coli*. *Annu. Rev. Genet.*, **18,** 31.

57. Sawyer, S. A. and Hartl, D. L. (1986) Distribution of transposable elements in prokaryotes. *Theor. Pop. Biol.*, **30,** 1.
58. Lawrence, J. G., Ochman, H., and Hartl, D. L. (1992) The evolution of insertion sequences within enteric bacteria. *Genetics*, **131,** 9.
59. Montgomery, E. A., Charlesworth, B., and Langley, C. H. (1987) A test for the role of natural selection in the stabilization of transposable element copy number in a population of *Drosophila melanogaster*. *Genet. Res.*, **49,** 31.
60. Skowronski, J. and Singer, M. F. (1986) The abundant LINE-1 family of repeated DNA sequences in mammals: genes and pseudogenes. *Cold Spring Harbor Symp. Quant. Biol.*, **51,** 457.
61. Spradling, A. C. (1986) P element-mediated transformation. In *Drosophila: A Practical Approach*. D. B. Roberts (ed.). IRL Press, Oxford, p. 175.
62. Ronsseray, S., Lehmann, M., and Periquet, G. (1989) Comparison of the regulation of P elements in M and M′ strains of *Drosophila melanogaster*. *Genet. Res.*, **54,** 13.
63. Burton, F. H., Loeb, D. D., Edgell, M. H., and Hutchinson, C. A., III (1991) L1 gene conversion or same site transposition. *Mol. Biol. Evol.*, **8,** 609.
64. Hu, W.-S. and Temin, H. M. (1990) Retroviral recombination and reverse transcription. *Science*, **250,** 1227.
65. Yokoyama, S. (1991) Molecular evolution of human immunodeficiency viruses and related retroviruses. In *Evolution at the Molecular Level*. R. K. Selander, A. C. Clark, and T. S. Whittam (eds). Sinauer, Sunderland, Mass. p. 96.
66. Dougherty, J. P. and Temin, H. (1986) High mutation rates of a spleen necrosis virus-based retrovirus vector. *Mol. Cell. Biol.*, **6,** 4387.
67. Muller, H. J. (1964) The relation of recombination to mutational advance. *Mutat. Res.*, **1,** 2.
68. Capy, P., Koga, A., David, J. R., and Hartl, D. L. (1992) Sequence analysis of active mariner elements in natural populations of *Drosophila simulans*. *Genetics*, **130,** 499.
69. Besansky, N. J. (1990) Evolution of the T1 retroposon family in the *Anopheles gambiae* complex. *Mol. Biol. Evol.*, **7,** 229.
70. Ullu, E. and Tschudi, C. (1984) Alu sequences are processed 7SL RNA genes. *Nature*, **312,** 171.
71. Quentin, Y. (1992) Origin of the Alu family: a family of Alu-like monomers gave birth to the left and right arms of the Alu elements. *Nucleic Acids Res.*, **20,** 3397.
72. Ohta, T. (1985) A model for duplicative transposition and gene conversion for repetitive DNA families. *Genetics*, **110,** 513.
73. Slatkin, M. (1985) Genetic differentiation of transposable elements under mutation and unbiased gene conversion. *Genetics*, **110,** 145.
74. Brookfield, J. F. Y. (1986) A model for DNA sequence evolution within transposable element families. *Genetics*, **112,** 393.
75. Ohta, T. (1986) Population genetics of an expanding family of mobile transposable elements. *Genetics*, **113,** 145.
76. Slatkin, M. and Hudson, R. R. (1991) Pairwise comparisons of mitochondrial DNA sequences in stable and exponentially-growing populations. *Genetics*, **129,** 555.
77. Kaplan, N. L. and Hudson, R. R. (1989) An evolutionary model for highly repeated interspersed DNA sequences. In *Mathematical Evolutionary Theory*. M. Feldman (ed.). Princeton University Press, Princeton, NJ, p. 301.
78. Britten, R. J., Baron, W. F., Stout, D. B., and Davidson, E. H. (1988) Sources and evolution of human Alu repeated sequences. *Proc. Natl Acad. Sci. USA*, **85,** 4770.

79. Britten, R. J., Stout, D. B., and Davidson, E. H. (1989) The current source of human Alu retroposons is a conserved gene shared with an old world monkey. *Proc. Natl Acad. Sci. USA*, **86,** 3718.
80. Deininger, P. L., Batzer, M. A., Hutchinson, C. A., III, and Edgell, M. H. (1992) Master genes in mammalian repetitive DNA amplification. *Trends Genet.*, **8,** 307.
81. Jurka, J. and Smith, T. (1988) A fundamental division in the Alu family of repeated sequences. *Proc. Natl Acad. Sci. USA*, **85,** 4775.
82. Leeflang, E. P., Liu, W.-M., Hashimoto, C., Chaudary, P. V., and Schmid, C. W. (1992) Phylogenetic evidence for multiple Alu source genes. *J. Mol. Evol.*, **35,** 7.
83. Matera, A. G., Hellmann, U., Hintz, M. F., and Schmid, C. W. (1990) Recently transposed Alu repeats result from multiple source genes. *Nucleic Acids Res.*, **18,** 6019.
84. Quentin, Y. (1988) The Alu family developed through successive waves of fixation closely connected with primate lineage history. *J. Mol. Evol.*, **27,** 194.
85. Shen, M. R., Batzer, M. A., and Deininger, P. L. (1991) Evolution of the master Alu gene(s). *J. Mol. Evol.*, **33,** 311.
86. Wallace, M. R., Andersen, L. B., Saulino, A. M., Gregory, P. E., Glover, T. W., and Collins, F. S. (1991) A *de novo* Alu insertion results in neurofibromatosis type I. *Nature*, **353,** 864.
87. Jurka, J. (1989) Subfamily structure and evolution of the human L1 family of repetitive sequences. *J. Mol. Biol.*, **29,** 496.
88. Hardies, S. C., Martin, S. L., Voliva, C. A., Hutchinson, C. A., III, and Edgell, M. H. (1986) An analysis of replacement and synonymous changes in the rodent L1 repeat family. *Mol. Biol. Evol.*, **3,** 109.
89. Martin, S. L., Voliva, C. F., Hardies, S. C., Edgell, M. H., and Hutchinson, C. A., III (1985) Tempo and mode of concerted evolution in the L1 repeat family of mice. *Mol. Biol. Evol.*, **2,** 127.
90. Demers, G. W., Matunis, M. J., and Hardison, R. C. (1989) The L1 family of long interspersed repetitive DNA in rabbits: sequence, copy number, conserved open reading frames, and similarity to keratin. *J. Mol. Evol.*, **29,** 3.
91. Scott, A. F., Schmeckpeper, B. J., Abdelrazik, M., Comey, C. T., O'Hara, B., Rossiter, J. P., Cooley, T., Heath, P., Smith, K. D., and Margolet, L. (1987) Origin of the human L1 elements: proposed progenitor genes deduced from a consensus DNA sequence. *Genomics*, **1,** 113.
92. Dover, G. A. (1982) Molecular drive: a cohesive mode of species evolution. *Nature*, **299,** 111.
93. Anxolabéhère, D., Kidwell, M. G., and Periquet, G. (1988) Molecular characteristics of diverse populations are consistent with the hypothesis of a recent invasion of *Drosophila melanogaster* by mobile *P* elements. *Mol. Biol. Evol.*, **5,** 252.
94. Daniels, S. B., Peterson, K. R., Strausbaugh, L. D., Kidwell, M. G., and Chovnick, A. (1990) Evidence for horizontal transmission of the *P* transposable element between *Drosophila* species. *Genetics*, **124,** 339.
95. Houck, M. A., Clark, J. B., Peterson, K. R., and Kidwell, M. G. (1991) Possible horizontal transfer of *Drosophila* genes by the mite *Proctolaelaps regalis*. *Science*, **253,** 1125.
96. McClure, M. (1991) Evolution of retroposons by acquisition or deletion of retrovirus-like genes. *Mol. Biol. Evol.*, **8,** 835.
97. Voytas, D. F., Cummings, M. P., Konieczny, A., Ausubel, F. M., and Rodermel,

S. R. (1992) *copia*-like retrotransposons are ubiquitous among plants. *Proc. Natl Acad. Sci. USA*, **89,** 7124.

98. Xiong, Y. and Eickbush, T. H. (1990) Origin and evolution of retroelements based upon their reverse transcriptase sequences. *EMBO J.*, **9,** 3353.
99. Gerasimova, T. I., Ilyin, Y. V., Mizrokhi, L. J., Semjonova, L. V., and Georgiev, G. P. (1984) Mobilization of the transposable element *mdg4* by hybrid dysgenesis generates a family of unstable cut mutations in *Drosophila melanogaster*. *Mol. Gen. Genet.*, **193,** 488.
100. Gerasimova, T. I., Matyunina, L. V., Ilyin, Y. V., and Georgiev, G. P. (1984) Simultaneous transposition of different mobile elements: relation to multiple mutagenesis in *Drosophila melanogaster*. *Mol. Gen. Genet.*, **194,** 517.
101. Lewis, A. P. and Brookfield, J. F. Y. (1987) Movement of *Drosophila melanogaster* transposable elements other than *P* elements in a P–M hybrid dysgenic cross. *Mol. Gen. Genet.*, **208,** 506.
102. Eggleston, W. B., Johnson-Schlitz, D. M., and Engels, W. R. (1988) P–M hybrid dysgenesis does not mobilise other transposable element families in *D. melanogaster*. *Nature*, **331,** 368.
103. Fayet, O., Ramond, P., Polard, P., Prère, M. F., and Chandler, M. (1990) Functional similarities between retroviruses and the IS3 family of bacterial insertion sequences? *Mol. Microbiol.*, **4,** 1771.
104. Kulkosky, J., Jones, K. S., Katz, R. A., Mack, J. P. G., and Skalka, A. M. (1992) Residues critical for retroviral integrative recombination in a region that is highly conserved among retroviral/retrotransposon integrases and bacterial insertion sequence transposases. *Mol. Cell. Biol.*, **12,** 2331.

7 | Retrons in bacteria

DONGBIN LIM, TANIA M. O. LIMA, and WERNER K. MAAS

1. Introduction

There are many genetic elements whose life-cycle depends on a reverse transcriptase (RT). These elements are generally called retro-elements and they are found among viruses, transposons, and plasmids. Such genetic elements are widely distributed in eukaryotes. Although there had been sporadic reports on the existence of RT in bacteria, it was not clear until recently that a similar element was present in prokaryotic organisms. A few years ago the production of an RT was clearly shown in two different kinds of bacteria, *Escherichia coli* and *Myxococcus xanthus*. Since then, it has been shown that there are RTs in several genera of bacteria.

The bacterial RTs have structural similarities to retroviral RTs. In contrast to retroviruses, the bacterial life-cycle does not depend on this enzyme and there is no known phenotype associated with its production, except that it is responsible for the synthesis of a multicopy single-stranded DNA, covalently linked to RNA and thus called msDNA. The genetic determinants for the synthesis of msDNA are associated with the gene for the RT and they are part of a genetic element referred to as a retron.

2. msDNA

2.1 Discovery

msDNA was discovered as a satellite band in polyacrylamide gel electrophoresis of total DNA from *M. xanthus*, a Gram-negative soil bacterium (1). When experiments were carried out to measure the genomic complexity by denaturation–renaturation kinetics of total DNA of *M. xanthus*, it was found that a significant portion of the DNA renatured quickly. When the total DNA of *M. xanthus* was subjected to polyacrylamide gel electrophoresis, a DNA satellite band of about 150 nucleotides in length was observed. The satellite band was shown to be a single-stranded DNA of 162 nucleotides. Since there were found to be 500 copies of this DNA molecule per cell, it was called multicopy single-stranded DNA, or msDNA (1a).

During the determination of the sequence of msDNA it was noticed that a large portion of the ^{32}P label at the 5′ end of msDNA was released by piperidine treatment, the final step of the Maxam–Gilbert sequencing reaction (2). This was interpreted to mean that the msDNA isolated after RNase treatment may contain a few alkali (piperidine)-labile residues at its 5′ terminus, such as ribonucleotides. In a series of elegant experiments, Furuichi *et al.* (2, 3) showed that the 162 nucleotide msDNA was derived from a branched RNA–DNA copolymer, obtained after treatment with RNase A. In the RNA–DNA compound, the 5′ end of msDNA is covalently linked by a 2′–5′ phosphodiester bond to an internal guanine residue of the RNA part (called msDNA-associated RNA or msdRNA), forming a branched nucleic acid. In the initial survey of the distribution of msDNA in various bacterial species, this type of nucleic acid was found only in myxobacteria and its relatives. It was thought, therefore, that its function might be related to the complex life cycle of these bacteria.

msDNA–RNA in *E. coli* was discovered serendipitously as an endogenous template/primer for retroviral RT. During studies on the regulation of arginine biosynthesis, Lim and Maas (4) performed a primer extension to measure the length of mRNA. Primer extension is usually done with an end-labelled primer. However, Lim and Maas used unlabelled primer, and added labelled nucleotide during the RT extension reaction. One might expect that with this protocol any DNA products formed by RT would be labelled. When the reaction product was analysed on a gel, a huge amount of labelled product was observed in the reactions with *E. coli* B. This product did not depend on the added primer and it was not formed in the reaction with *E. coli* K12. The extended product was shown to be a covalently linked RNA–DNA compound. From this observation it was concluded that *E. coli* B produces an endogenous template/primer for retroviral RT. Further analysis showed that this DNA–RNA compound was structurally similar to the previously characterized msDNA–RNA compound of *M. xanthus*. Analysis showed that in the primer extensions carried out in the original experiment, the Moloney murine leukaemia virus (MoMLV) RT extended the msDNA to the branch position of msdRNA using the msdRNA as a template (Fig. 1).

After the discovery of msDNA in *E. coli* B, it was shown that there are other msDNAs in different strains of *E. coli* (11, 20). Therefore, a system for nomenclature was introduced based on the length of msDNA and the strain that produces it. According to this system, the *M. xanthus* msDNA, which is 162 nucleotides long, is called msDNA Mx-162. Similarly, the msDNA of *E. coli* B is called msDNA Ec-86, since it is 86 nucleotides long.

2.2 Structure of msDNA

Structures of two representative msDNA–RNAs, one from *M. xanthus* (Mx-162) and the other from *E. coli* B (Ec-86), are shown in Fig. 2. At present nine different msDNA–RNA compounds are known. There are two different types of msDNA in myxobacteria. One of these, Mx-62, is prevalent in all isolates of *M. xanthus*. Furthermore, msDNA identical to or related to Mx-162 is present in other species of myxobacteria. However, most msDNAs in *E. coli* are not related in their primary sequence. Analysis of several msDNAs showed that, in spite of the high divergence of nucleotide sequences among msDNAs, there are common features in

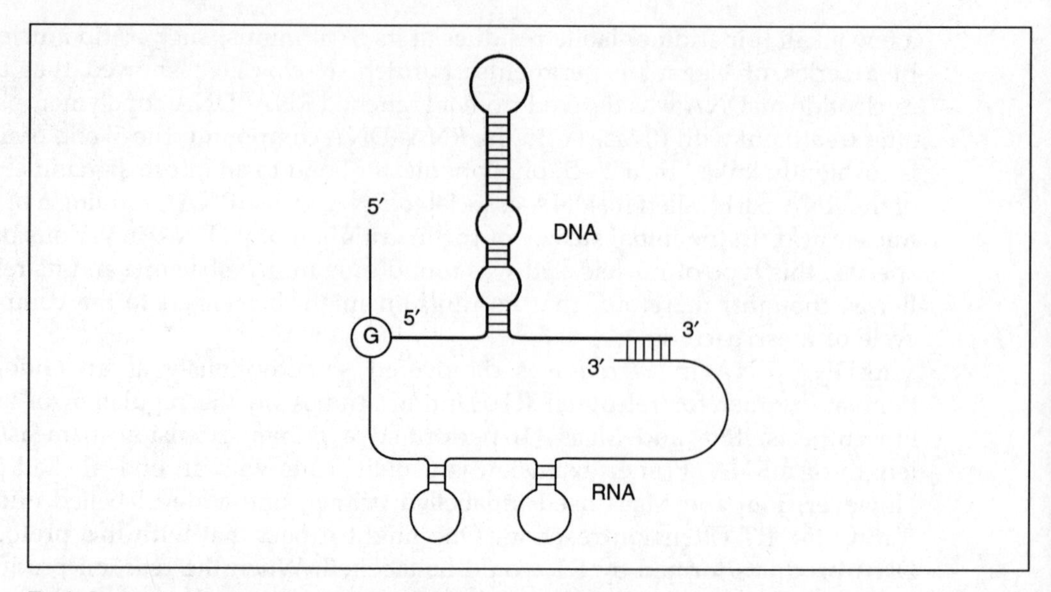

Fig. 1 Generalized version of an msDNA–RNA compound. Variable features include: the distance from the 5′ end of msdRNA to the branched guanine residue, the length of msDNA, the length of the DNA–RNA hybrid region at the 3′ ends, and the length of msdRNA. Base-paired regions are indicated by cross bars and the DNA part is shown in bold.

msDNA–RNA molecules of *E. coli* and myxobacteria (6). These common characteristics are as follows.

1. The 5′ end of msDNA is covalently linked to an internal guanine residue of msdRNA by a 2′–5′ phosphodiester bond, forming a branched DNA–RNA copolymer. The branched residue of msdRNA is always a guanine. The 5′ arm in the RNA molecule (upstream of the branched guanine) is part of inverted repeats present within the coding region of the msDNA-msdRNA.
2. The 3′ ends of msDNA and msdRNA form a stable DNA–RNA hybrid, containing 5–8 bases.
3. Both msDNA and RNA form a secondary stable stem–loop structure due to the presence of complementary nucleotides within the msDNA sequence.

The conservation of the msDNA structure during the evolution of bacteria was clearly demonstrated when msDNAs from two myxobacteria were compared. When an msDNA from *M. xanthus* was compared with one from *Stigmatella aurantiaca*, a myxobacterium related to *M. xanthus*, there was found to be 81 per cent identity in the sequences of msDNA and msdRNA (7). There are 21 base substitutions between these two msDNAs and almost all of these substitutions are such that the secondary structure of the compound is conserved. Such conservation of secondary structure during evolution and the presence of common features in unrelated msDNA–RNAs could be the result of structural requirements for the biosynthesis or for the functions of msDNA.

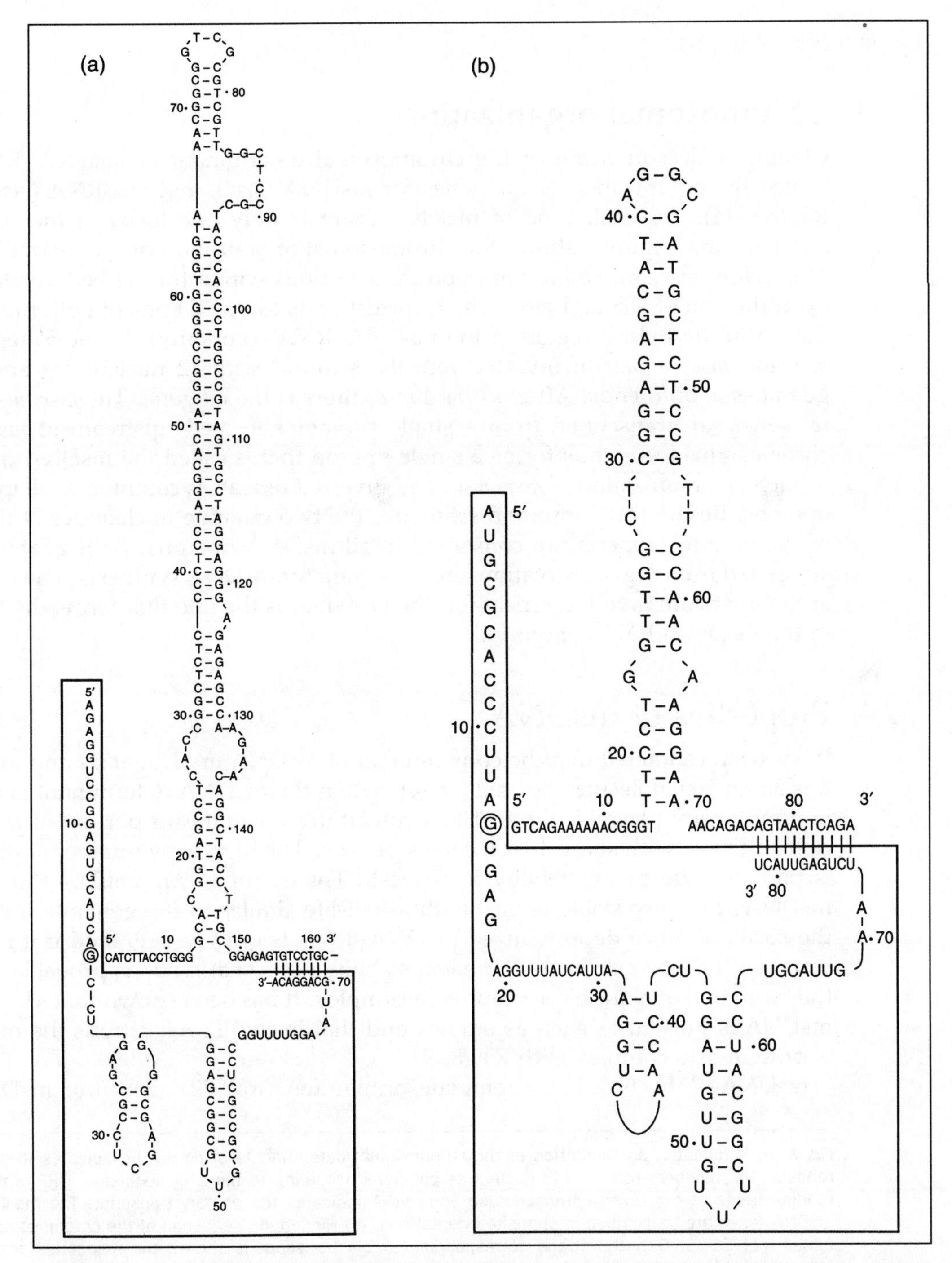

Fig. 2 Structures of Mx-162, one of the msDNAs found in *M. xanthus* isolates (a), and of Ec-86, the msDNA found in *E. coli* (b). Sequences within the boxed region are RNA, and those outside are DNA. The branching ribonucleotide is rG (represented as Ⓖ, from which a 2′–5′ phosphodiester linkage to the first deoxynucleotide, a dC in Mx-162 and a dG in Ec-86, is formed. The numbers refer to nucleotide positions, numbered separately for the RNA and the DNA parts, starting from the 5′ ends of msDNA and msdRNA. Complementary base-pairings are shown by bars between two bases.

2.3 Chromosomal organization

Cloning and sequencing of the chromosomal determinant of msDNA–RNA revealed the organization of the genes for msDNA (*msd*), and msdRNA (*msr*), and RT (*ret*) (4). For each type of msDNA there is only one locus in the bacterial chromosome. Figure 3 shows the chromosomal organization of the msDNA locus of *E. coli* B. *msr* and *msd* are in opposite directions with a transcribed overlap of 8 bp at the end of *msr* and *msd*, which corresponds to the 3′ ends of both molecules that form the hybrid region of the msDNA–RNA compound. In the 5′ region of *msr* and *msd*, a pair of inverted repeats is found with 12 nucleotides and with guanines at both ends. After *msr* and *msd*, there is the *ret* gene. The *msr*, *msd*, and *ret* genes are transcribed from a single promoter located upstream of *msr*. The chromosomal locus thus forms a single operon that is called the msDNA operon.

Such an organization of *msr*, *msd*, and inverted repeats is common to all msDNA operons. Beside this conserved structure, the two guanine nucleotides at the end of the inverted repeats are conserved in all msDNA operons. Both guanines, as suggested from the conservation, are important for msDNA synthesis. The guanine at the upstream inverted repeat (on the *msr* side) is the one that forms the branch in the msDNA–RNA compound.

2.4 Properties of msDNA

It has been estimated that the copy number of msDNA in *M. xanthus* and in *E. coli* B is about 500 molecules per cell (1a, 4). When the msDNA determinant is cloned in a multicopy plasmid, the msDNA constitutes a significant portion of the total cellular DNA, with about 10 000 copies per cell. The high copy number of msDNA is partially due to its stability in the cell. For example, *M. xanthus* and *E. coli* msDNAs are very stable *in vivo* with a half-life similar to the generation time of the bacteria. Since deproteinized msDNA–RNA is quickly degraded if it is incubated with *E. coli* extract, the unusual stability observed *in vivo* is probably due to the formation of a nucleic acid–protein complex. It has been shown that *M. xanthus* msDNA is present as such a complex and that in two *E. coli* retrons the msDNA is present as a complex with RT (8, 9).

msDNA–RNA is a good template/primer for viral RTs. *In vitro*, msDNA is

Fig. 3 (a) Schematic representation of the chromosomal determinants of Ec-86. IR indicates the inverted repeats. The numbers refer to the nucleotide positions according to the DNA sequence. ORF is the open reading frame of the reverse transcriptase and mRNA indicates the primary transcript. The msdRNA and msDNA genes are transcribed in opposite orientations. (b) Nucleotide sequence of the chromosomal determinants of Ec-86. Only the strand corresponding to the transcript is shown; the respective amino acid sequence is shown in the three-letter code. Nucleotides are numbered starting from the first base observed in the msdRNA. The msdRNA coding region is overlined and the region that encodes msDNA is underlined. The msDNA sequence is complementary to the sequence shown. Inverted repeats (IRs) are indicated by heavy dashed lines. The circled G at position 14 is the branched guanylate of msdRNA in the msDNA–RNA compound, from which reverse transciption is primed. The rG at position 156 (marked with ▼) is another conserved guanine at the other end of the IRs.

(a)

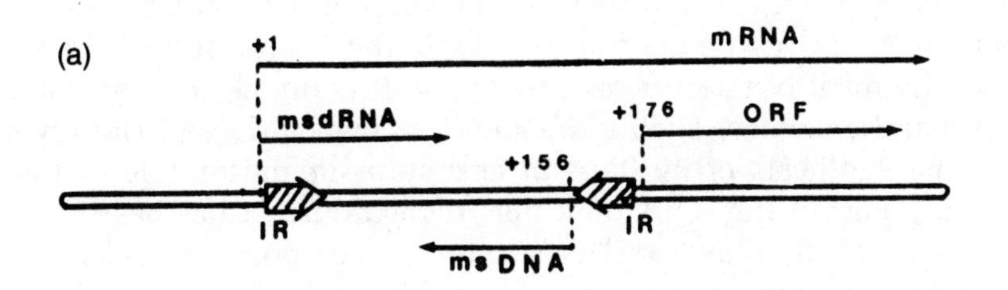

(b)

```
-371  TGGCATCTATTAAGAAGGTTAGGAAAGAAAATAAAGTATCAAAAGATATTGGAAATATAT

-311  TATACGCAGAGCGTTTCTATTGCCTTGTATCTATTTACTGGATAGTGTCAACTACCGCAC

-251  ACTGTGTGAACTAGCTTTTAAAGCGATAAAGCAAGATGATGTTTTATCTAAAATTATTGT

-191  TAGATCCGTTGTTTCTCGTCTAATAAATGAACGAAAAATACTTCAAATGACTGATGGTTA

-131  TCAGGTCACTGCTTTGGGGGCTAGCTATGTTAGGAGCGTCTTTGATAGAAAGACACTTGA

 -71  CCGATTGCGGCTTGAGATTATGAATTTTGAAAACCGTAGAAAATCAACATTTAACTATGA
                +1
                .|-|--------------------msdRNA---------------------------
 -11  TAAGATTCCGTATGCGCACCCTTAGCGAGAGGTTTATCATTAAGGTCAACCTCTGGATGT
                IR ============
      ---------------------------------->.
  49  TGTTTCGGCATCCTGCATTGAATCTGAGTTACTGTCTGTTTTCCTTGTTGGAACGGAGAG
                            <----------------------------------------

 109  CATCGCCTGATGCTCTCCGAGCCAACCAGGAAACCCGTTTTTTCTGACGTAAGGGTGCGC
      ----------------------msDNA------------------------| ==========

 169  AACTTTCATGAAATCCGCTGAATATTTGAACACTTTTAGATTGAGAAATCTCGGCCTACC
      = IR   MetLysSerAlaGluTyrLeuAsnThrPheArgLeuArgAsnLeuGlyLeuPr

 229  TGTCATGAACAATTTGCATGACATGTCTAAGGCGACTCGCATATCTGTTGAAACACTTCG
      oValMetAsnAsnLeuHisAspMetSerLysAlaThrArgIleSerValGluThrLeuAr

 289  GTTGTTAATCTATACAGCTGATTTTCGCTATAGGATCTACACTGTAGAAAAGAAAGGCCC
      gLeuLeuIleTyrThrAlaAspPheArgTyrArgIleTyrThrValGluLysLysGlyPr

 349  AGAGAAGAGAATGAGAACCATTTACCAACCTTCTCGAGAACTTAAAGCCTTACAAGGATG
      oGluLysArgMetArgThrIleTyrGlnProSerArgGluLeuLysAlaLeuGlnGlyTr

 409  GGTTCTACGTAACATTTTAGATAAACTGTCGTCATCTCCTTTTTCTATTGGATTTGAAAA
      pValLeuArgAsnIleLeuAspLysLeuSerSerSerProPheSerIleGlyPheGluLy

 469  GCACCAATCTATTTTGAATAATGCTACCCCGCATATTGGGGCAAACTTTATACTGAATAT
      sHisGlnSerIleLeuAsnAsnAlaThrProHisIleGlyAlaAsnPheIleLeuAsnIl

 529  TGATTTGGAGGATTTTTTCCCAAGTTTAACTGCTAACAAAGTTTTTGGAGTGTTCCATTC
      eAspLeuGluAspPhePheProSerLeuThrAlaAsnLysValPheGlyValPheHisSe

 589  TCTTGGTTATAATCGACTAATATCTTCAGTTTTGACAAAAATATGTTGTTATAAAAATCT
      rLeuGlyTyrAsnArgLeuIleSerSerValLeuThrLysIleCysCysTyrLysAsnLe

 649  GCTACCACAAGGTGCTCCATCATCACCTAAATTAGCTAATCTAATATGTTCTAAACTTGA
      uLeuProGlnGlyAlaProSerSerProLysLeuAlaAsnLeuIleCysSerLysLeuAs

 709  TTATCGTATTCAGGGTTATGCAGGTAGTCGGGGCTTGATATATACGAGATATGCCGATGA
      pTyrArgIleGlnGlyTyrAlaGlySerArgGlyLeuIleTyrThrArgTyrAlaAspAs

 769  TCTCACCTTATCTGCACAGTCTATGAAAAAGGTTGTTAAAGCACGTGATTTTTTATTTTC
      pLeuThrLeuSerAlaGlnSerMetLysLysValValLysAlaArgAspPheLeuPheSe

 829  TATAATCCCAAGTGAAGGATTGGTTATTAACTCAAAAAAAACTTGTATTAGTGGGCCTCG
      rIleIleProSerGluGlyLeuValIleAsnSerLysLysThrCysIleSerGlyProAr

 889  TAGTCAGAGGAAAGTTACAGGTTTAGTTATTTCACAAGAGAAAGTTGGGATAGGTAGAGA
      gSerGlnArgLysValThrGlyLeuValIleSerGlnGluLysValGlyIleGlyArgGl

 949  AAAATATAAAGAAATTAGAGCAAAGATACATCATATATTTTGCGGTAAGTCTTCTGAGAT
      uLysTyrLysGluIleArgAlaLysIleHisHisIlePheCysGlyLysSerSerGluIl

1009  AGAACACGTTAGGGGATGGTTGTCATTTATTTTAAGTGTGGATTCAAAAAGCCATAGGAG
      eGluHisValArgGlyTrpLeuSerPheIleLeuSerValAspSerLysSerHisArgAr

1069  ATTAATAACTTATATTAGCAAATTAGAAAAAAAATATGGAAAGAACCCTTTAAATAAAGC
      gLeuIleThrTyrIleSerLysLeuGluLysLysTyrGlyLysAsnProLeuAsnLysAl

1129  GAAGACCTAATGGTCTTCGTTTTAAAACTAAAGCTCATAGGTTGAAAAATTGAGCACTTC
      aLysThr

1189  TTCGTCCAACCAGTTATTTAGTTCCTGCAATCGTTTCTGCAG
```

efficiently extended to the branch position of msdRNA by bacterial and retroviral RTs (4, 8). In such extension reactions, the 3′ end of msDNA serves as a primer and the msdRNA serves as a template. It is not clear if the bacterial RT, like the retroviral enzymes, is able to extend its own msDNA. The reverse transcription for the synthesis of msDNA *in vivo* stops in the middle of the template RNA, leaving part of the RNA molecule as msdRNA. It has been shown that bacterial RT can extend its own msDNA to the branch point of msdRNA (4, 8). However, one must exercise caution in taking this as a property of bacterial RT, because, unlike the extension by the retroviral reverse transcriptase, only a small portion of msDNA is extended by its own enzyme. Moreover, purified msDNA–RNA has not been tested as a substrate. It is possible that in the reported results only intermediates of msDNA synthesis, which are present in the substrate preparation, are extended beyond the normal termination site. Extension to the branch point can be observed *in vivo*, especially in RNase H mutant strains of *E. coli*, in which msDNA synthesis appears to stop at random at several points, leaving a series of intermediates either smaller or bigger than the wild-type msDNA.

2.5 Distribution

msDNA is found in several species of myxobacteria and in enteric bacteria, such as *Proteus mirabilis*, *Klebsiella pneumoniae*, and *Rhizobium trifolii* (10). An identical or a very closely related msDNA has been detected in all strains of *M. xanthus* isolated independently from different places. A similar msDNA was identified in various strains of myxobacteria, such as *Stigmatella aurantiaca*, *Myxococcus coralloides*, *Cystobacter virolaceous*, *Cystobacter ferrugines*, and *Nannocystis exedens* (7). *M. xanthus* (Mx-162) and *S. aurantiaca* (Sa-163) msDNAs share about 81 per cent sequence identity; there are five insertions, four deletions, and 21 base substitutions between these two msDNA–RNAs. Interestingly, almost all base substitutions occurred in such a way that the stem–loop structure of msDNA and msdRNA was conserved.

That msDNA is present in all isolates of *M. xanthus*, and that various myxobacterial species produce a related msDNA, strongly suggest that the ancestral myxobacteria had an msDNA and that all msDNAs in myxobacterial species were vertically inherited from this common ancestor. Since myxobacteria are believed to be very ancient organisms, it is reasonable to assume that the myxobacterial retron is very old.

The distribution of *E. coli* msDNA is quite different from that in myxobacteria. In contrast to *M. xanthus*, most *E. coli* isolates, like *E. coli* K-12, do not produce msDNA; only a small portion of natural isolates produce this DNA. In addition, different *E. coli* isolates have different msDNAs. From the clonal analysis of *E. coli* isolates it was shown that, with few exceptions, a particular msDNA is associated with a particular clone of *E. coli* (11). The data suggest that msDNA elements are foreign to *E. coli* and, in contrast to myxobacteria that they only became associated with this bacterium fairly recently; and also that the acquisition of different retrons by different clones of *E. coli* occurred independently. For example, the location of msDNA on the chromosome is different in several msDNA-producing strains, suggesting independent insertions. Furthermore, two different retrons, Ec-86 of *E. coli* B and Ec-67 of a clinical strain, are present in the same location with the same flanking sequences (12). This can only have occurred by independent integration into these two strains. The retron Ec-107 (13), a short 1.3 kb fragment, was integrated into the genome of some wild-type *E. coli* isolates by replacing a 34 bp intergenic sequence between the *pyrE* and *ttk* genes, at position 82 min on the chromosome; it is found in isolates representing three divergent branches of the *E. coli* phylogenetic tree and it is the most frequently found retron in wild *E. coli* strains (13). This could be evidence for the horizontal transmission of the retron between different lines of *E. coli*, but it is unclear if there is a retron-specific mechanism of transmission; there are no other genes in this retron besides *msr*, *msd*, and *ret*, and it is doubtful whether the Ec-107 retron itself is mobile.

At present, the source or the mechanism of integration of *E. coli* retrons is obscure, even though it is clear that, once the msDNA element had become successfully associated with the *E. coli* genome, it was stably inherited in these strains.

3. Reverse transcriptase

3.1 Discovery

Bacterial RT was discovered as an enzyme responsible for the synthesis of msDNA. The presence of a *trans*-acting factor for msDNA synthesis was suggested by the decrease of msDNA synthesis by deletion and insertion mutants generated at the downstream region of retron Mx-162. The responsible protein was identified after cloning of the *E. coli* B msDNA genetic determinant (4). From the determination of the minimal region required for the biosynthesis of msDNA Ec-86, it was found that an open reading frame (ORF) located close to *msd* was required for the production of msDNA in *E. coli* K12. This ORF encodes a predicted protein of 320 amino acids, which has significant homology with the amino acid sequence of retroviral RTs (4). Subsequently it was shown that the protein encoded by the ORF has RT activity *in vitro*. Similarly, an ORF product responsible for the biosynthesis of msDNA Ec-67 was identified and its RT activity was confirmed (14). The enzyme has since been found in all retrons, its genes usually being located downstream of *msr* and *msd*. The distance between the RT ORF and the *msd* gene varies from 19 to 77 bp. However, in retron Ec-73 an ORF (ORF 316) with unknown function is present between *msd* and the RT ORF. In retron EC-86, *ret* is transcribed from a promoter located in the upstream region of *msd* and is part of the msDNA operon.

3.2 Structure

In general retroviral RTs are composed of more than 550 amino acids and show two enzymatic activities: RNA-dependent DNA polymerase activity and RNase H activity. These two activities are localized at two distinct domains of a single polypeptide. The RT domain is composed of about 250 amino acids and is located in the N-terminal part. The RNase H domain is in the C-terminal part, and is about

120 amino acids long. These two domains are connected by a tether of variable length.

The size of the polypeptide encoded by ORFs in retrons varies from 316 amino acids (in Ec-73) to 586 amino acids (Ec-67). Amino acid sequence comparison shows that the homologous region among bacterial RTs spans about 250 amino acids. The RTs from myxobacterial retrons are longer than those present in *E. coli*. Along with the RT domain, they have a long N-terminal domain with an unknown function. All *E. coli* RTs, except that from Ec-67, are considerably shorter than retroviral ones. They are composed of about 320 amino acids; of these, about 250 are well aligned with retroviral RTs (6). From structural and genetic studies, it is known that this is an RNA-dependent DNA polymerase domain. Thus, most *E. coli* RTs are simple enzymes with only the RT domain. The RT of Ec-67 is considerably longer than that of other *E. coli* RTs (586 amino acids) and its C-terminal region can be aligned with the sequence of RNase H. However, it is not known if this similarity to RNase H is significant, since retron Ec-67 requires host RNase H activity for msDNA synthesis.

Bacterial RTs share some features with retroviral RTs, so they may be related evolutionarily (Fig. 4). All the known RTs have a highly conserved YXDD sequence (in all bacterial RTs the X position is an alanine). The sequence similarities among bacterial RTs are few. Of the 250 residues that constitute the RT domain in these enzymes, only 33 identical amino acids are found; these 33 are also common in eukaryotic RTs (6).

Myxococcus xanthus produces two different msDNAs (Mx-162 and Mx-65) with two different RTs. With respect to organization and amino acid sequence, these two RTs are more closely related to each other than to any *E. coli* RT. For instance, both myxobacterial RTs have a long N-terminal domain which is not observed in *E. coli* RTs (6). However, even in the RT domains, the homology between these two proteins is less than 50 per cent and the two msDNA synthesis systems appear to be independent so that the RT of each retron cannot complement the msDNA synthesis of the other retron (6). The sequence similarity between *E. coli* RTs is not as high, with only about 30 per cent identity in the amino acid sequence. The *E. coli* hosts of such retrons share a very high degree of identity in DNA and protein sequences. This shows that the evolutionary history of retrons is independent of that of the host organisms. It appears that such divergent retrons evolved outside *E. coli* and that the retrons were integrated into the *E. coli* genome recently on the evolutionary time scale. Further support for this hypothesis is obtained when the codon usages of bacterial RTs are compared. In myxobacteria, the codon usage is very similar to that of other host genes; however, among *E. coli* RTs, it is substantially different from that of the host genes (6).

3.3 Properties

In spite of earlier claims of RT activity in *E. coli* (15), it was previously believed that the results obtained in those experiments were due to a DNA polymerase I

```
                     *       *       *     *    *   *
Mx162 RWFAFHREVD TATHYVSWTI PKRDGS--KR TITSPKPELK AAQRWVLSNV VERLPVH--- -----GAAHG 226
Mx65  RHYSIHRPRE RVRHYVTFAV PKRSGG--VR LLHAPKRRLK ALQRRMLALL VSKLPVS--- -----PQAHG 195
Ec67  FTLNVLYRIG SDNQYTQFTI PKKGKG--VR TISAPTDRLK DIQRRICDLL SDCRDEIFAI RKISNNYSFG 96
Ec86  VETLRLLTYT ADFKYRIYTV EKKGDEKRMR TIYQPSRELK ALQGWVLRNI LDKLSSS--- -----PFSIG 95
Ec73  TKGFASEVMR SPEPPKKWDF AKKKGG--MR TIYHPSSKVK LIQYWLMNVF -SKLPMH--- -----NAAYA 73
Ec79  PDFDVLLKSR PATHYKVYKI PKRTIG--YR IIAQPTPRVK AIQRDIIEIL KQHTHIH--- -----DAATA 73

      *    **  * *  *            ** ** * *
Mx162 FVAGRSILTN ALAH--QGAD VVVKVDLKDF FPSVTWRRVK GLLRKGGLRE GTSTLLSLLS TEAAVQFRGK 294
Mx65  FVPGRSIKTG AAPH--VGRR VVLKLDLKDF FPSVTFARVR GLLLALGYGY PVAATLAVLM TESERQPVEL 263
Ec67  FERGKSIILN AYKH--RGKQ IILNIDLKDF FESFNFGRVR GYFLSNQDFL LNPVVATTLA ---------- 154
Ec86  FEKHQSILNN ATPH--IGAN FILNIDLEDF FPSLTANKVF GVFHSLGYNR LISSVLT--- ---------- 150
Ec73  FVKNRSIKSN ALLHAESKNK YYVKIDLKDF FPSIKFTDFE YAFTRYRDRI EFTTEYDLEL LQLIKT---- 139
Ec79  YVDGKNILDN AKIH--QSSV YLLKLDLVNF FNKITPELLF KALARQKVDI SDTNKNLLKQ FCFWNRT--- 138

                     * * *  **   *       *                 *  *****  *
Mx162 LL---HVAKG PRALPQGAPT SPGITNALCL KLDKRLSALA ---KRLGFTY TRYADDLTFS WTKAKQPKPR 361
Mx65  EGILFHVPVG PRVCVQGAPT SPALCNAVLL RLDRRLAGLA ---RRYGYTY TRYADDLTFS GDDVTA---- 326
Ec67  -----KAACY NGTLPQGSPC SPIISNLICN IMDMRLAKLA ---KKYGCTY SRYADDITIS TNKNTFPLEN 216
Ec86  -----KICCY KNLLPQGAPS SPKLANLICS KLDYRIQGYA ---GSRGLIY TRYADDLTLS AQSMKK---- 208
Ec73  -----ICFIS DSTLPIGFPT SPLIANFVAR ELDEKLDQKL NAIDKLNATY TRYADDIIVS TNMKGA---- 200
Ec79  -----KRKNG ALVLSVGAPS SPFISNIVMS SFDEEISSFC ---KENKISY SRYADDLTFS TNERDV---- 196

                                                   ***
Mx162 RTQRPPVAVL LSRVQEVVEA EGFRVHPDKT RVARK-GTRQ RVTGLVV---( 78 )----485 amino acids
Mx65  ------LERV RALAARYVQE EGFEVNREKT RVQRR-GGAQ RVTGVTV---( 61 )----427 amino acids
Ec67  ATVQPEGVVL GKVLVKEIEN SGFEINDSKT RLTYK-TSRQ EVTGLTV---( 324)----586 amino acids
Ec86  ------VVKA RDFLFSIIPS EGLVINSKKT CISGP-RSQR KVTGLVI---( 72 )----320 amino acids
Ec73  -----SKLIL DCFKRTMKEI GPDFKINIKK FKICSAGGSI VVTGLKV---( 73 )----316 amino acids
Ec79  ----LGLAHQ KVKTTLIRFF GTRIIINNNK IVYSSKAHNR HVTGVTL---( 73 )----313 amino acids
```

Fig. 4 Alignments of the amino acid sequences of the RTs from bacterial retrons. Standard single-letter abbreviations are used to designate amino acids. The stars indicate the amino acids which are common to most retrons shown in this figure. The numbers in parentheses indicate the remaining number of amino acids found in each retron and which are not shown. The homology between bacterial RTs is not high, as can be seen from the number of common amino acids. Note the RT motif YXDD found in all eukaryotic, viral, and bacterial RTs. In bacterial retrons, the X is always an alanine. Amino acids which are underlined are the ones which are conserved in eukaryotic RTs.

activity, since it has been shown that this enzyme has intrinsic RT activity. Therefore, to demonstrate that an RNA-dependent DNA polymerase activity was encoded in the retron, a DNA polymerase I-deficient strain (*polA*) was used (14). In this experiment [α-^{32}P]dGTP was incorporated when a poly(rC) oligomer was used as a template-primer system. In addition, partially purified RT from Ec-86 was able to extend msDNA to the branched rG residue of msdRNA, in a manner similar to what is obtained when retroviral (Mo-MuLV) RT is used.

The RTs found in bacteria resemble retroviral RTs in their ability to use different templates and primers *in vitro*. Purified preparations of bacterial RT are able to synthesize a cDNA by extending the 3′ end of a synthetic DNA primer annealed to the 3′ end of a 5S rRNA serving as template, and can synthesize DNA using a DNA template annealed to a DNA or RNA primer.

It seems that RT is present in the bacterial cell as a complex with msDNA. When extracts prepared from Ec-67 or Ec-86 are passed through columns, RT is co-purified with msDNA as a high molecular weight complex (6). msDNA from *M.*

xanthus also appears to exist as a complex with proteins. It is not clear whether other proteins are also part of this complex but it seems likely that the host RNase H is part of it.

In most cases, the 2′-OH priming reaction appears to be specific for each RT; for example, msDNA Ec-67 can by synthesized by the RT from its own retron but not by an RT from retron Ec-73 and vice versa (10a). Specificity may depend on the secondary structure of the precursor RNA in the vicinity of the branched rG residue, which is unique for individual msdRNAs (10b). However, some heterologous RTs can synthesize msDNA from another retron, although much less efficiently than the cognate msDNA: the Ec-67 RT can synthesize *E. coli* B msDNA (but not the reverse) and the Ec-74 RT can synthesize unbranched msDNA from Ec-79 (see below). This generally stringent requirement of a specific RT for the priming reaction for each msDNA is in sharp contrast to the fact that bacterial RTs can use different templates (RNA or DNA) or primers for the chain elongation reaction *in vitro*. The process is in this way similar to retroviral cDNA synthesis, since retroviral RTs bind to specific tRNAs for priming in a highly specific reaction.

4. Biosynthesis of branched msDNA

4.1 *Trans*-acting genes for msDNA synthesis

The first indication of the presence of *trans*-acting genes for the msDNA synthesis came from mutational studies on retron Mx-162 in *M. xanthus*. When Dhundale *et al.* (16) introduced transposon insertions in the downstream region of *msr*, these mutants produced significantly reduced amounts of msDNA. This suggested that genes other than *msr* and *msd* are required for msDNA synthesis (4). When retron Ec-86 was cloned from *E. coli* B, it was determined that a 1.2 kb fragment contained the minimal region required for msDNA synthesis. Besides the genes for msDNA and its associated RNA, this fragment contains a single ORF in the downstream region of *msr*. The amino acid sequence encoded by this ORF is similar to that of retroviral RTs and the RT activity of the protein encoded by this ORF was confirmed *in vitro*. Subsequently, a similar RT ORF has been found in all msDNA synthesis systems (10). Without exception, the three genes for msDNA synthesis are located in the chromosome as a single copy in the order *msr*, *msd*, and *ret*. In *E. coli* B, as described above, these genes form a single operon with a promoter located upstream of *msr*. This appears to be true in all retrons. To produce msDNA, however, *msr–msd* and *ret* do not have to be transcribed into a single RNA molecule; for example, msDNA is synthesized if *msr–msd* and *ret* are cloned into two separate plasmids (19).

By screening about 10 000 mutants generated by Tn*5* insertion, three chromosomal mutants defective in msDNA synthesis were isolated (T. Lima, unpublished results). Analysis of these mutants showed that in all three the Tn*5* transposon had inserted into the chromosomal gene for RNase H (*rnh*). Thus the RNase H activity required in msDNA synthesis comes from the host. Considering the large number of mutants screened, RNase H may be the only host protein functioning in msDNA synthesis.

4.2 RNA template

In principle the template for msDNA could be chromosomal DNA or a transcribed RNA. In the initial model suggested for msDNA synthesis the template was chromosomal DNA and msDNA was made by a cellular DNA polymerase using RNA as a primer. The first indication of an RNA template came from the effects of antibiotics on msDNA synthesis. Furuichi *et al.* (3) showed that rifampicin but not nalidixic acid inhibits msDNA production in *M. xanthus*. Rifampicin inhibits RNA production by interfering with RNA polymerase and nalidixic acid interferes with DNA replication by inhibiting DNA gyrase. Also, these authors identified an RNA transcript covering the *msr* and *msd* genes. On the basis of these facts, Lampson *et al.* (14) proposed a model for RT-dependent msDNA synthesis, which turned out to be essentially correct.

4.3 Mechanism of msDNA synthesis

A model for msDNA synthesis is schematically drawn in Fig. 5. A long mRNA covering *msr*, *msd*, and *ret* is transcribed from a promoter located upstream of *msr*. RT is produced from the translation of the *ret* region of this primary transcript. The *msr* and *msd* parts of the transcript serve as a precursor for msdRNA and as a template for msDNA synthesis. They fold into a particular structure because of the presence of a pair of inverted repeats (IRs) at the ends of *msr* and *msd*; such a structure is recognized by the RT which copies the *msd* region from the precursor RNA. The 2′-OH of the guanine residue at the end of the inverted repeat serves as a primer for reverse transcription. With the progression of DNA synthesis, the RNA in the RNA–DNA hybrid region is digested by RNase H. When the RT reaches a specific point of the template RNA, the reaction terminates, leaving approximately half of the template RNA as msdRNA. Since the primer is the 2′-OH of the internal guanine in the template, the DNA product is covalently linked to RNA (3).

Various points of this model have been tested, such as structural requirements of the template and the precursor RNA (10a, 10b, 17). It has been shown that the guanine residue at the branch point is absolutely required for msDNA synthesis. It has also been shown that the inverted repeats are important for the efficient synthesis of msDNA. A base change in the IR region had little effect if the IR structure was reconstructed by a secondary mutation at the complementary base of the IR (17). The msDNA was synthesized as a branched DNA–RNA, showing that the branch formation is not a postsynthetic reaction. Since the guanine at the branch point is required for the initiation of reverse transcription, it is believed that the 2′-OH of this guanine is the primer for reverse transcription.

The involvement of RNase H in msDNA synthesis was shown in *M. xanthus* and

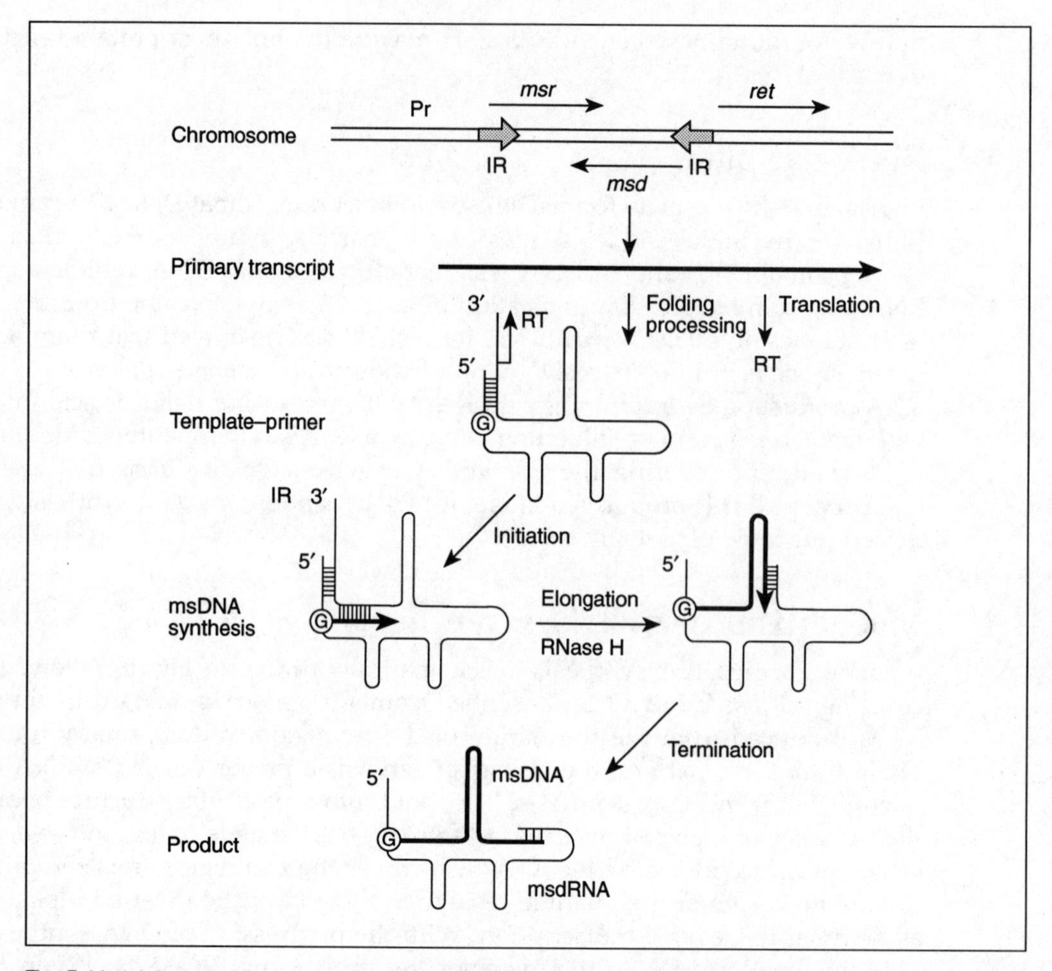

Fig. 5 Model for the biosynthesis of an msDNA–RNA compound. The organization of the chromosomal determinants is shown in the top line, from which the primary RNA transcript is synthesized. Pr indicates the location of the promoter of the retron; *msr* codes for msdRNA, *msd* for msDNA, and *ret* for RT. The shaded arrows labelled IR indicate the inverted repeats. The primary transcript starts from the promoter and covers *msr*, *msd*, and *ret*. The RT ORF is translated. Part of the primary transcript, not including the RT portion, folds and forms a secondary structure, which serves as template and primer for the RT to synthesize msDNA. The double-stranded region created by the inverted repeats, which presumably directs the initiation process to start at the circled G, is indicated by cross bars. The 2′-OH group of the circled G serves as a primer for the RT, which then elongates the DNA molecule with concomitant digestion of the RNA by RNase H. The reverse transcription stops at a very specific site of the template, leaving about half of it as msdRNA.

E. coli (18). When chain-terminating dideoxynucleotides were added to the cell-free synthesis, DNA intermediates of various lengths were obtained. However, the combined length of the DNA and RNA components of these intermediates was unchanged regardless of the length of the DNA components. This was interpreted to mean that reverse transcription proceeds with concomitant digestion of the RNA in the DNA–RNA hybrid.

4.4 Template structure

Some structural features of the RNA template seem to be essential for msDNA synthesis; such features are conserved in all msDNAs isolated so far, despite their sequence and size diversities. The stem structure immediately upstream of the branched G residue is essential (6). When base mismatches are introduced in the stem structure formed by a pair of IRs located immediately upstream of the branched rG residue, msDNA synthesis is almost completely blocked. If, however, additional mutations are introduced on the other strand to restore the complementary base pairing, msDNA synthesis is recovered, indicating that there is no specific sequence requirement immediately upstream of the branched rG residue, except to maintain the secondary structure.

The G residue in the RNA template from which msDNA synthesis is primed from its 2′ position cannot be replaced by another residue (C or A (6)). There is no stringent requirement for the 5′ end of msDNA, which is linked to the branched rG, in contrast to the stringent requirement for the branched rG residue that serves as a primer.

4.5 Priming

The priming reaction (the addition of the first deoxynucleoside base to the RNA template) has been demonstrated in a cell-free system (10) with total RNA isolated from cells containing only the *msr–msd* region of retron Ec-67, with no *ret* gene, and partially purified RT from the same retron. In this cell-free system, only dTTP (but not dATP, dGTP, or dCTP) was incorporated into a 132 base precursor RNA, yielding a 133 base compound. This specific dT incorporation could be altered to dA or dC by simply substituting a dT or dG for the dA residue in the template RNA corresponding to the first base, indicating that the priming reaction is a specific template-directed event. Once the priming reaction was completed, RT was able to elongate the DNA chain further, yielding the same sequence as msDNA Ec-67.

At present, the exact biochemical mechanism by which msDNA synthesis is primed through this guanine residue is unknown. The structure around the branched G residue is highly conserved and so is the stem structure of the IRs.

4.6 Termination and host factors

Even though the synthesis of msDNA *in vivo* terminates at a very precise position of the template, leaving approximately half of it in the form of msdRNA, *in vitro* experiments have shown that msDNA can serve as a substrate for retroviral or bacterial RTs that can further extend the DNA strand up to the branched G residue (6). Since the msDNA molecules *in vivo* seem to be present in a complex with proteins, it seems likely that a protein factor might block the RT activity at a very specific point during the synthesis. Such a factor is not encoded by the retron itself, since it contains only the genes for msdRNA, msDNA, and RT. Presumably it would be a host-encoded protein, probably bound to the stem–loop structure of msdRNA, preventing the RT from further extension.

In fact, the msDNA molecule is frequently co-purified with the RT in a high molecular weight complex, and msDNA has been shown to be associated with the enzyme in the cell. RT presumably remains bound to the msDNA–RNA hybrid after the synthesis is completed (8, 9).

However, the postulated terminator protein has not yet been isolated and the mechanism of termination is still poorly understood. Recent data suggest the involvement of RNase H in this process (Lim and Lima, unpublished). We have isolated three independent chromosomal mutants which are defective in msDNA synthesis; all of these are RNase H mutants. Studies of msDNA synthesis in an *rnh* mutant showed that in the absence of RNase H activity the termination of reverse transcription is abnormal. In an *rnh* strain, reverse transcription stops at many sites, before or after the termination site, resulting in DNA fragments of various lengths. We interpret this observation to mean that for correct termination, the RT should recognize the secondary and tertiary structures formed by the DNA and RNA. If RNA is not digested by RNase H, as in an *rnh* mutant, the synthesized DNA would be in the DNA–RNA hybrid state and, therefore, would not be able to form the canonical secondary structure necessary for correct termination.

5. Unbranched msDNA

Lim (19) described an msDNA isolated from the *E. coli* clinical strain 161 that, unlike the retrons previously analysed, is present in the cell, free of RNA, with a monophosphate at its 5′ end. This msDNA is 79 nucleotides long and is derived from a branched DNA–RNA hybrid, whose structure is similar to that of the other retrons described so far.

The chromosomal organization of this retron is the same as that of other retrons, with *msr*, *msd*, and *ret*, suggesting that the basic mechanism of reverse transcription in strain 161 is similar to that of the other retrons. IR sequences are also present at both ends and the position of these IRs relative to the branched guanine and to the 5′ end of DNA fits completely with the structural organization of the *msd–msr* region of other retrons.

A model for the biosynthesis of msDNA Ec-79 assumes that the RT produces a msDNA linked to RNA by a mechanism similar to that already described above. However, unlike other msDNA–RNA compounds, the DNA–RNA hybrid in this strain is further processed to a linear msDNA without RNA. This processing reaction is an endonucleolytic cleavage of the DNA moiety of the DNA–RNA compound. The substrate for such a cleavage, the branched DNA–RNA compound, apparently has an unusual structure, as judged from its electrophoretic mobility. This processing of the DNA–RNA hybrid occurs very efficiently *in vivo* and the major species present in the cell is the unbranched linear single-stranded

DNA. The cleavage is not random but cuts the phosphodiester bond of the DNA strand precisely between the fourth and the fifth positions of msDNA.

It is not clear how this cleavage reaction is performed. It is very unlikely that this hypothetical enzyme is encoded by the *E. coli* chromosome since such msDNA cleavage is not observed in previously characterized retrons. Rather, we propose that the DNA cleavage activity is an inherent property of some retrons. Since the RT is the only protein encoded by the minimum region required for the synthesis and cleavage of this msDNA, either the RT or the msDNA–RNA compound must have the endonucleolytic activity. When the amino acid sequence of the RT ORF was compared with that of other RTs producing the branched msDNA–RNA compound, no unusual domains specific for this RT were found. Therefore, it is not clear whether or not the DNA cleavage is carried out by the RT. Since the RNA-linked msDNA molecule has unusual structural properties, as indicated by its abnormal gel mobility, it is also possible that the hybrid itself may be a catalyst. In this case, the RT protein may play a structural, rather than a catalytic role, in the cleavage reaction.

Previously it had been shown (20) that of five different msdNAs of *E. coli* clinical isolates, two (strains 161 and 110) were poorly extended by RT. It has now been demonstrated that both of them are free of RNA. This occurrence of two unbranched msDNAs out of five suggests that retrons producing msDNA unlinked to msdRNA are not rare. The function of such cleavage is a matter of speculation. In the case of the yeast retrotransposon *Ty1*, a debranching enzyme is required for its efficient transposition (21). If msDNA functions as a primer for reverse transcription, as frequently suggested, the msDNA free of RNA would be a better primer. However, there is no evidence for reverse transcription of any other genes in bacteria, besides *msd*.

6. Retrons

6.1 Structure and distribution

The genetic determinants responsible for the synthesis of msDNA are discrete elements, even though at least one cellular component, RNase H, is also necessary for their synthesis (6). They comprise short DNA sequences of 2–3 kb in the bacterial chromosome and the three genes belong to a single operon. Retrons can thus be considered as primitive retro-elements encoding only one protein, the RT, in contrast to other retro-elements, such as retroviruses and retrotransposons, which also encode an integrase. Temin (22) proposed the name *retron* to describe such a genetic element.

Retrons can be classified into two categories: those present in myxobacteria and those in *E. coli*. Myxobacteria retrons are very similar and are present in all the isolates studied so far. It is likely that they are derived from a common progenitor retron, which would have been acquired before the myxobacteria species diverged. It is possible that these retrons are older than any other known eukaryotic retroelement and that they later acquired other retro-element features, such as *gag* and *int*, resulting in non-LTR (long terminal repeat) retrotransposons, which were proposed as the oldest group of retro-elements by Xiong and Eickbush (23). The *E. coli* retrons were probably acquired more recently, after different *E. coli* lineages had been established (6). Only a minor fraction of *E. coli* wild strains (approximately 13 per cent) contain retrons. Retrons are not restricted to clinical strains; strains isolated from healthy people as well as wild strains isolated from their natural habitats contain msDNA (24). *Escherichia coli* retrons are highly divergent and are integrated into different sites on the *E. coli* chromosome; codon usages in the RT genes are substantially different from the average codon usage in the *E. coli* genome. Inouye and Inouye (10) propose that before these retrons were integrated into the *E. coli* genome, they were probably replicated like retroviruses and retro-transposons, in a highly error-prone system, using RTs for their reproduction.

6.2 Retronphage and integration site

Some retrons (Ec-73, Ec-86, and Ec-67) are parts of prophages (12, 25, 26). Retron Ec-73 is part of a prophage related to bacteriophage P4 and is integrated into the *selC* gene (for selenocysteyl-tRNA) of the *E. coli* chromosome. It was found that it can be excised from the chromosome upon infection with helper phage P2 and it can produce an infectious P4-like virion (27). This virion, called retronphage ϕR73, is able to integrate into the *selC* gene of a new host cell, thus enabling the newly formed lysogen to produce msDNA Ec-73.

Retron Ec-67 is part of a 34 kb foreign DNA fragment integrated in the *E. coli* chromosome and flanked by 26 bp direct repeats (25). It is a prophage related to phage 186, and Ec-86 is part of a similar prophage (12). These retrons somehow became integrated into the phage genomes and are probably able to transfer from one cell to another by using bacteriophages as vectors.

Recent data show that Ec-86 is part of a defective retronphage (ϕR86) (28). Under conditions that activate the RecA proteins, involved in the SOS response, this retronphage can be induced, killing the host cell. Induction of ϕR86 results in inhibition of host DNA replication before cell death. A retronphage 'killer' gene, ORF336, causes similar effects without SOS induction when overexpressed from a plasmid. It seems probable that the RecA730 protein induces ϕR86 by promoting autodigestion of its repressor, as in lambda prophage induction. Interestingly, ϕR86 is not induced by UV-activated RecA protein. Some ORFs in ϕR86 have considerable homology to phage 186 genes, suggesting that this phage might have acquired the retron by an unknown mechanism and later became integrated in the *E. coli* chromosome.

7. Evolutionary considerations and perspective

The common mode of information transfer in biological systems is DNA to RNA to protein. Almost 20 years ago it was established that certain viruses use information transfer from RNA to DNA catalysed by RT. It was later found that RT and the products of its action are widely distributed in eukaryotes (29). The discovery of bacterial RTs demonstrates that this enzyme is not unique to eukaryotes, but is more ancient than the separation between prokaryotes and eukaryotes.

Some evolutionary biologists have suggested that there was an earlier RNA world, so that a RT activity would have been needed for a template-based switch to a DNA-based life. If the first genes were indeed in the form of RNA, reverse transcription would have been a crucial early step in the evolution of life (30, 31). The discovery of RT in bacteria revealed that RTs have been present in bacteria for a long time on the evolutionary time-scale. msDNAs and their RTs can thus be seen as a 'missing link'. Their occurrence in Archaebacteria still has to be assessed; such a study could provide more insights into the origins of life.

If the argument for the antiquity of the myxobacterial RT gene is accepted, then the precursor of this gene was the ancestor of all present day retro-elements. Thus, the path of evolution went from retrons, to retroposons with the ability to transpose, to retrotransposons with LTRs, to retroviruses with the ability to form virus particles, to pararetroviruses which have lost the ability to integrate and transpose, and, perhaps, to some circular DNA viruses (22).

Is there a cellular role for RTs? There have been several suggestions that reverse transcription is somehow advantageous to its hosts (32, 33). However, all previously described RTs seem primarily to benefit the RT gene itself or the retro-element that encodes it, rather than the host containing the RT gene. Thus, there are many retro-elements which are parasitic and whose main physiological role relative to cellular organisms is their own self-reproduction. Nevertheless, the myxobacterial and *E. coli* RT genes appear to be single-copy, indicating that their function is not to amplify the retron relative to other bacterial genes; that is, the retron is not 'selfish' DNA. Most bacteria appear to thrive without these loci, but selective advantages or disadvantages might be subtle. For example, as proposed by Varmus (34), the synthesis of many copies of msDNA might impair growth, or the RTs might be used to duplicate or repair genes by copying mRNAs into DNA.

One thing that is clear in considering possible functions of the bacterial retrons is that the synthesis of msDNA from an RNA template ensures that the sequence of msDNA is hypervariable; this could be a clue to its function. The question is why this ability to synthesize multiple copies of a small hypervariable piece of DNA has been conserved in some bacteria through evolution. Maybe msDNA could be a convenient primer for the copying of useful mRNAs and subsequent gene conversion (35). Recently it has been found that the presence of retrons Ec-86 and Ec-79 leads to increased mutation frequencies (36). It was suggested that this increase is due to the binding of components of the mismatch repair system (37) to mismatches in the step loop structure of msDNA. In support of this suggestion it was shown that retron Ec-74, which has no mismatches in the stem-loop structure of its msDNA, is not mutagenic (36).

In any case, the msDNA-synthesizing enzymes extend the striking diversity of priming mechanisms used by RTs. Intact or fragmented host tRNAs, oligoribonucleotides generated by RNase H, and even proteins have been identified as primers for reverse transcription (34). The discovery that msDNA synthesis is primed from an internal branched ribonucleotide raises questions of evolutionary antecedence and replication strategy. The first DNA polymerases could have used sites on template molecules as primers, rather than independent molecules. It is now clear that DNA synthesis can assume a variety of forms; such diversity creates even more expectations for further discoveries.

Acknowledgements

The work by Dongbin Lim was supported by the Ministry of Education of Korea, Genetic Engineering Research Program, and by the Korean Science and Engineering Foundation, KOSEF 931-0500-008-2.

References

1. Yee, T. and Inouye, M. (1981) Reexamination of the genome size of Myxobacteria, including the use of a new method for genome size analysis. *J. Bacteriol.*, **145,** 1257.

1a. Yee, T., Furuichi, T., Inouye, S., and Inouye, M. (1984) Multicopy single-DNA isolated from a Gram-negative bacterium, *Myxococcus xanthus*. *Cell*, **38**, 203.

2. Furuichi, T., Dhundale, A., Inouye, M., and Inouye, S. (1987) Branched RNA covalently linked to the 5′ end of a single-stranded DNA in *Stigmatella aurantiaca*: structure of msDNA. *Cell*, **48,** 47.

3. Furuichi, T., Dhundale, A., Inouye, M., and Inouye, S. (1987) Biosynthesis and structure of stable branched RNA covalently linked to the 5′ end of multicopy single-stranded DNA of *Stigmatella aurantiaca*. *Cell*, **48,** 55.

4. Lim, D. and Maas, W. K. (1989) Reverse transcriptase-dependent synthesis of a covalently linked, branched DNA–RNA compound in *E. coli* B. *Cell*, **56,** 891.

5. Lim, D. and Maas, W. K. (1989) Reverse transcriptase in bacteria. *Mol. Microbiol.*, **3,** 1141.

6. Inouye, M. and Inouye, S. (1991) msDNA and bacterial reverse transcriptase. *Annu. Rev. Microbiol.*, **45,** 163.

7. Dhundale, A., Furuichi, T., Inouye, S., and Inouye, M. (1985) Distribution of multicopy single-stranded DNA among myxobacteria and related species. *J. Bacteriol.*, **164,** 914.

8. Lampson, B. C., Viswanathan, M., Inouye, M., and Inouye, S. (1990) Reverse transcriptase from *Escherichia coli* exists as a complex with msDNA and is able to synthesize double-stranded DNA. *J. Biol. Chem.*, **265,** 8490.

9. Viswanathan, M., Inouye, M., and Inouye, S. (1989) *Myxococcus xanthus* msDNA—Mx162—exists as a complex with proteins. *J. Biol. Chem.*, **264,** 13665.

10. Inouye, M. and Inouye, S. (1992) Retrons and multicopy single-stranded DNA. *J. Bacteriol.*, **174,** 2419.

10a. Shimamoto, T., Hsu, M., Inouye, S., and Inouye, M. (1993) Reverse transcriptases from bacterial retrons require specific secondary structures at the 5′-end of the template for the cDNA priming reaction. *J. Biol. Chem.*, **268,** 2684.

10b. Shimada, M., Inouye, S., and Inouye, M. (1994) Requirements of the secondary structures in the primary transcript for multicopy single-stranded DNA synthesis by reverse transcriptase from bacterial retron-Ec107. *J. Biol. Chem.*, **269,** 14553.

11. Sun, J., Herzer, P. J., Weinstein, M. P., Lampson, B. C., Inouye, M., and Inouye, S. (1989) Extensive diversity of branched RNA-linked multicopy single-stranded DNAs in clinical strains of *Escherichia coli*. *Proc. Natl Acad. Sci. USA*, **86**, 7208.
12. Lim, D. (1991) Structure of two retrons of *Escherichia coli* and their comon chromosomal insertion site. *Mol. Microbiol.*, **5**, 1863.
13. Herzer, P. J., Inouye, S., and Inouye, M. (1992) Retron Ec107 is inserted into the *Escherichia coli* genome by replacing a palindromic 34 bp intergenic sequence. *Mol. Microbiol.*, **6**, 345.
14. Lampson, B. C., Sun, J., Hsu, M. H., Vallejo-Ramirez, J., Inouye, S., and Inouye, M. (1989) Reverse transcriptase in a clinical strain of *Escherichia coli*: production of branched RNA-linked msDNA. *Science*, **243**, 1033.
15. Vorob'eva, N. V., Nebrat, L. T., Potapov, V. A., Romashchenko, A. G., Salganik, R. I., and Yushkova, L. F. (1982) Reverse transcription of heterologous RNA with the help of RNA-dependent DNA polymerase of *Escherichia coli*. *Mol. Biol.*, **16**, 770.
16. Dhundale, A., Furuichi, T., Inouye, M., and Inouye, S. (1988) Mutations that affect production of branched RNA-linked msDNA in *Myxococcus xanthus*. *J. Bacteriol.*, **170**, 5620.
17. Hsu, M. Y., Inouye, S., and Inouye, M. (1989) Structural requirements of the RNA precursor for the biosynthesis of the branched RNA-linked multicopy single-stranded DNA of *Myxococcus xanthus*. *J. Biol. Chem.*, **264**, 6214.
18. Lampson, B. C., Inouye, M., and Inouye, S. (1989) Reverse transcriptase with concomitant ribonuclease H activity in the cell-free synthesis of branched RNA-linked msDNA of *Myxococcus xanthus*. *Cell*, **56**, 701.
19. Lim, D. (1992) Structure and biosynthesis of unbranched multicopy single-stranded DNA by reverse transcriptase in a clinical *Escherichia coli* isolate. *Mol. Microbiol.*, **6**, 3531.
20. Lim, D., Gomes, T. A. T., and Maas, W. K. (1990) Distribution of msDNAs among serotypes of enteropathogenic *Escherichia coli* strains. *Mol. Microbiol.*, **4**, 1711.
21. Chapman, K. B. and Boeke, J. D. (1991) Isolation and characterization of the gene encoding yeast debranching enzyme. *Cell*, **65**, 483.
22. Temin, H. M. (1989) Retrons in bacteria. *Nature*, **339**, 254.
23. Xiong, Y. and Eickbush, T. H. (1990) Origin and evolution of retroelements based upon their reverse transcriptase sequences. *EMBO J.*, **9**, 3353.
24. Herzer, P. J., Inouye, S., Inouye, M., and Whittam, T. S. (1990) Phylogenetic distribution of branched RNA-linked multicopy single-stranded DNA among natural isolates of *Escherichia coli*. *J. Bacteriol.*, **172**, 6175.
25. Hsu, M. Y., Inouye, M., and Inouye, S. (1990) Retron for the 67-base multicopy single-stranded DNA from *Escherichia coli*: a potential transposable element encoding both reverse transcriptase and Dam methylase functions. *Proc. Natl Acad. Sci. USA*, **87**, 9454.
26. Sun, J., Inouye, M., and Inouye, S. (1991) Association of a retroelement with a P4-like cryptic prophage (retronphage ϕR73) integrated into the selenocystyl tRNA gene of *Escherichia coli*. *J. Bacteriol.*, **173**, 4171.
27. Inouye, S., Sunshine, M. G., Six, E. W., and Inouye, M. (1991) Retronphage ϕR73: an *E. coli* phage that contains a retroelement and integrates into a tRNA gene. *Science*, **252**, 969.
28. Kirchner, J., Lim, D., Witkin, E. M., Gravey, N., and Roegner-Maniscalco, V. (1992) An SOS-inducible defective retronphage (ϕR86) in *Escherichia coli* strain B. *Mol. Microbiol.*, **6**, 2815.
29. Weiner, A. M., Deininger, P. L., and Efstratiadis, A. (1986) Nonviral retrotransposons:

genes: pseudogenes, and transposable elements generated by the reverse flow of genetic information. *Annu. Rev. Biochem.*, **55,** 631.
30. Joyce, G. F. (1989) RNA evolution and the origins of life. *Nature*, **338,** 217.
31. Waldrop, M. M. (1989) Did life really start out in an RNA world? *Science*, **246,** 1248.
32. Greider, C. W. and Blackburn, E. H. (1989) A telomeric sequence in the RNA of *Tetrahymena* telomerase required for telomere repeat synthesis. *Nature*, **337,** 331.
33. Temin, H. M. (1970) Malignant transformation of cells by viruses. *Perspect. Biol. Med.*, **14,** 11.
34. Varmus, H. E. (1989) Reverse transcription in bacteria. *Cell*, **56,** 721.
35. Cairns, J., Overbaugh, J., and Miller, S. (1988) The origin of mutants. *Nature*, **335,** 142.
36. Maas, W. K., Wang, C., Lima, T., Zubay, G., and Lim, D. (1994) Multicopy single-stranded DNAs with mismatched base pairs are mutagenic in *Escherichia coli*. *Mol. Microbiol.*, **14**(3), 437.
37. Modrich, P. (1991) Mechanisms and biological effects of mismatch repair. *Annu. Rev. Genet.* **25,** 299–53.

Index

Bold page numbers denote reference to figures.